Selected Titles in This Series

Sources of Hyperbolic Geometry

HISTORY OF MATHEMATICS
VOLUME 10

Sources of Hyperbolic Geometry

John Stillwell

AMERICAN MATHEMATICAL SOCIETY
LONDON MATHEMATICAL SOCIETY

1991 *Mathematics Subject Classification*. Primary 51-03;
Secondary 01A55, 53Axx, 51A05, 30F35.

Library of Congress Cataloging-in-Publication Data

Stillwell, John.
Sources of hyperbolic geometry / John C. Stillwell.
p. cm. — (History of mathematics; v. 10)
Includes bibliographical references (p. –) and index.
ISBN 0-8218-0529-0 (hardcover : alk. paper)
1. Geometry, Hyperbolic—History—Sources. I. Title. II. Series.
QA685.S83 1996
516.9—dc20 96-3894
CIP

∞ The paper used in this book is acid-free and falls within the guidelines
established to ensure permanence and durability.
The London Mathematical Society is incorporated under royal Charter
and is registered with the Charity Commissioners.

10 9 8 7 6 5 4 3 2 1 99 98 97 96

Preface

Hyperbolic geometry is the Cinderella story of mathematics. Rejected and hidden while her two sisters (spherical and euclidean geometry) hogged the limelight, hyperbolic geometry was eventually rescued and emerged to outshine them both. The first part of this saga – how Bolyai and Lobachevsky laboured in vain to win recognition for their subject – is well known, and English translations of the key documents are available in Bonola's classic *Non-Euclidean Geometry.* However, the turning point of the story has not been documented in English until now.

Beltrami came to the rescue of hyperbolic geometry in 1868 by interpreting it on a surface of constant negative curvature. By giving a concrete meaning to the hyperbolic plane, he put Bolyai's and Lobachevsky's work on a sound logical foundation for the first time, and showed that it was a part of classical differential geometry. This was quickly followed by interpretations in projective geometry by Klein in 1871, and in the complex numbers by Poincaré in 1882.

Hyperbolic geometry had arrived, and with Poincaré it joined the mainstream of mathematics. He used it immediately in differential equations, complex analysis, and number theory, and its place has been secure in these disciplines ever since. He also began to use it in low-dimensional topology, an idea kept alive by a handful of topologists until the spectacular blossoming of this field under Thurston in the late 1970s. Now, hyperbolic geometry is the generic geometry in dimensions 2 and 3.

Alongside these developments, there has been increased interest in the work of Beltrami, Klein, and Poincaré that made it all possible. I have had a steady stream of requests for the translations of Beltrami I produced in 1982, so I was delighted to be approached by Jim Stasheff with a proposal for a volume in the AMS-LMS history of mathematics series. I am also grateful to Bill Reynolds for his interest, and for help with the hyperboloid model, and to Abe Shenitzer for correcting a number of embarrassing errors in the Beltrami and Klein translations.

Clayton, Victoria, Australia John Stillwell

Contents

Translator's Introduction

Beltrami's *Essay on the interpretation of noneuclidean geometry*

It is generally known that Eugenio Beltrami's 1868 paper *Saggio di Interpretazione della Geometria Non-euclidea* (*Giornale di Mathematiche* **VI** (1868), 284-312) was the first to offer a concrete interpretation of hyperbolic geometry, by interpreting "straight lines" as geodesics on a surface of constant negative curvature. However, popular accounts of Beltrami's work suggest that it was confined to the *pseudosphere*, the surface of revolution generated by the tractrix (Figure 1.1), which models only a part of the hyperbolic plane. This supposed deficiency in Beltrami's work has been underlined by translating the "Saggio" in his title as "attempt" (Struik [1961], p. 152) or "Versuch" (Klein [1928], p. 285). More competent linguists than myself have advised me that "essay" is more likely than "attempt," but in any case, if Beltrami says "attempt" it is probably out of modesty, judging by the humble tone of his introduction, rather than lack of confidence.

The pseudosphere was already known to be a surface of constant negative

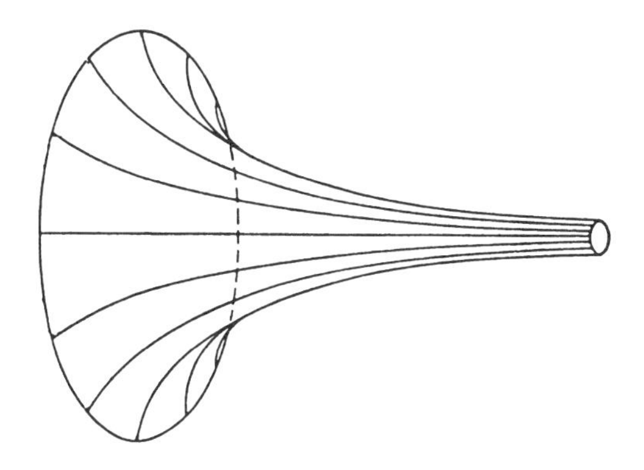

Figure 1.1: The pseudosphere

curvature, and Gauss's student Minding had investigated its geometry in the 1830s. Minding [1840] in fact came close to realising hyperbolic geometry on the pseudosphere by showing that the formulas for trigonometry on the pseudosphere were the same as those discovered by Bolyai and Lobachevsky for the hyperbolic plane.

However, Beltrami is well aware that the pseudosphere cannot be interpreted as a plane, because it has a boundary curve, and also because it is not simply connected. It is topologically a cylinder, not a plane. Beltrami achieves simple connectivity by considering instead the universal cover, a surface wrapped infinitely many times around the pseudosphere. This covering surface is still not a complete plane, because it has a boundary curve – the *horocycle* or "circle with centre at infinity" that covers the boundary curve of the pseudosphere. He implicitly removes the boundary curve by observing that it can be redrawn away from the boundary, as an arbitrary circle with the same centre at infinity. This is because the horocyclic disc represented by the covering surface is congruent to all horocyclic discs (in his terminology "any horocycle on the surface can be superimposed on any other"). Thus the universal cover of the pseudosphere can represent an arbitrary horocyclic disc, and hence an arbitrarily large portion of the hyperbolic plane.

The covering of the pseudosphere is actually the second universal covering mentioned in the paper. Beltrami introduces the idea of covering with another surface of constant negative curvature, defined by his equation (14). It does not have a name as far as I know, but is similar in shape to the *catenoid*, which Beltrami knew as one of the classical *minimal* surfaces – those for which the sum of the principal curvatures is zero. The Gaussian curvature of minimal surfaces is generally negative but not constant. Minding [1838] made the remarkable discovery that the catenoid can be cut and bent into a piece of another classical minimal surface, the *helicoid* (Figure 1.2, from Struik [1961], p. 121). Conversely, the helicoid is the universal cover of the catenoid. It is plausible that Beltrami noticed this covering first, then realised that a similar construction applied to his surface (14) of constant negative curvature. This would account for his describing its cover as "the limit of a helicoid."

Beltrami's reason for covering the pseudosphere is to find a "real" surface that models the whole hyperbolic plane – one that lies in ordinary space and has the ordinary notion of length of curves. As he explains in the introduction to his essay, he wishes to find a "real substrate" for the doctrine of Lobachevsky, "rather than admit the necessity for a new order of entities and concepts." The construction shows more sophistication than most commentators give Beltrami credit for (in particular, the universal cover is the first one I know in the literature), but it is not his main achievement. It is merely a prelude to something much more modern.

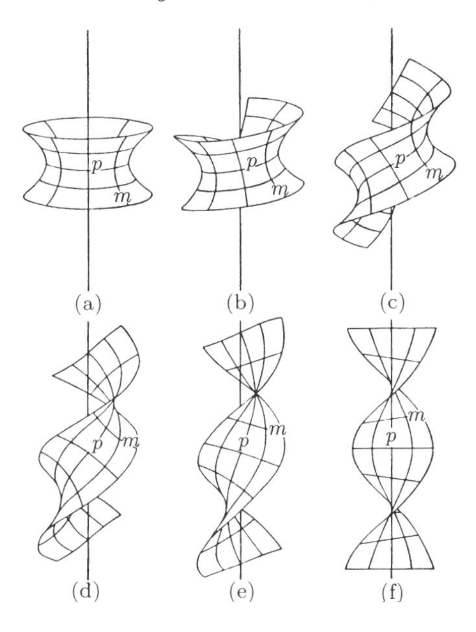

Figure 1.2: Bending the catenoid to the helicoid (Struik [1961], p. 121; reprinted with permission from D. J. Struik).

The surface we call the pseudosphere is not given a name by Beltrami. It is only a special case of what he calls a "pseudospherical surface," an *arbitrary* surface of constant negative curvature, *given abstractly* by the formula for its line element. This is what really interests him. Liouville [1850] had already found an abstract *simply connected* surface of constant negative curvature by transforming coordinates on the pseudosphere, namely, the upper half plane $y > 0$ with line element $\sqrt{dx^2 + dy^2}/y$. Beltrami finds a similar surface and notices for the first time that it is a model of Bolyai-Lobachevsky geometry, with geodesics playing the role of lines. Moreover, he makes another choice of coordinates, which gives two advantages:

- The "plane" is the open unit disc, hence bounded, and the "lines" are ordinary euclidean line segments – chords of the disc.

- The isometries of this "plane" are projective transformations of the euclidean plane that map the unit disc onto itself. (They are recognised in Note II of the essay, where Beltrami calls them by the old name "homographic correspondences.")

This is what later came to be called the *Klein disc model*, or the *projective model* of the hyperbolic plane.

Beltrami was originally led to this model by asking: which surfaces admit a map into the euclidean plane sending their geodesics to straight lines? It was known that the (hemi)sphere admits such a map (central projection onto a tangent plane), and Beltrami [1865] proved the remarkable result

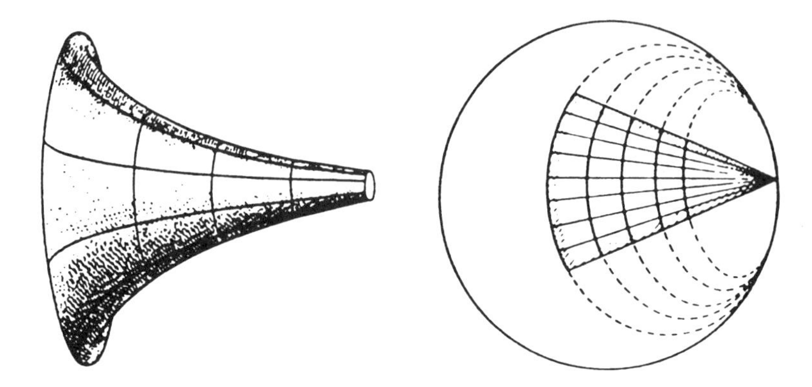

Figure 1.3: Map of the pseudosphere in the Klein model (Klein [1928], p. 286; reprinted with permission from Springer-Verlag).

that the only such surfaces are those of constant curvature. He outlines the proof in Note I to the essay. Applying his construction to the pseudosphere gives a horocyclic sector of the unit disc, with geodesics on the pseudosphere mapped to line segments within the sector. These segments have a natural extension to the whole interior of the disc, which is how Beltrami arrives at the Klein model. Figure 1.3, which is taken from Klein [1928], p. 286, shows the horocyclic sector corresponding to one turn around the pseudosphere. The dotted ellipses are the horocycles that wrap infinitely often around the various closed curves drawn on the surface. Any of them can in fact be "superimposed on any other" by a projective transformation of the disc.

This model first appeared in Cayley [1859], though as part of projective geometry and without reference to hyperbolic geometry. Klein [1928], p. 304, gives credit to Cayley, but not to Beltrami, and says that he (Klein) realised in 1869 that projective geometry contains a model of the hyperbolic plane. It seems strange that Klein could not see his own model in the *Saggio* paper, but could recognise a model of Poincaré in Beltrami's next paper, *Teoria fondamental degli spazii di curvatura costante* (Klein [1928], p. 300, Fricke and Klein [1897], p. 32). However, Beltrami's model was not genuinely projective from Klein's point of view, because it assumed the euclidean plane. (See the translation of Klein [1871], later in this volume.)

Today, we can see the *Saggio* as a crucial step in the acceptance of hyperbolic geometry, yet it was not universally accepted at the time. Gray [1992] tells us that Beltrami even delayed its publication for a year, because of criticisms by Cremona. This helps to explain the change to a more abstract outlook at the end, as well as Beltrami's mention of his forthcoming paper on spaces of constant curvature. Evidently, in the intervening time, he became convinced that Riemann's abstract approach to curvature in n-dimensional space (first published in 1867) was the way forward. As he says in a footnote near the end of his main text:

> In the present work we have been interested mainly in offering a concrete counterpart of abstract geometry; however, we do not wish to omit a declaration that the validity of the new order of concepts does not depend on the possibility of such a counterpart.

In conclusion, I would like to recommend the paper by Milnor [1982], which first alerted me to Beltrami's work. My geometric introduction to some of Poincaré's papers (Poincaré [1985]) was influenced by Milnor, and part of it has been used again here.

References

E. Beltrami [1865], Risoluzione del problema: "Riportare i punti di una superficie sopra un piano in modo che le linee geodetiche vengano rappresentate da linee rette." *Ann. Mat. pura appl.*, ser. 1, **7**, 185–204.

A. Cayley [1859], A sixth memoir upon quantics. *Mathematical Papers*, Vol. 2, 561–606.

R. Fricke and F. Klein [1897], *Vorlesungen über die Theorie der automorphen Funktionen*, Vol. I, Teubner, Stuttgart.

J. Gray [1992], Poincaré and Klein – Groups and Geometries. *1830-1930: A Century of Geometry* (Ed. L. Boi, D. Flament and J.-M. Salanskis), Springer-Verlag, 35–44.

F. Klein [1871] Über die sogenannte Nicht-Euklidische Geometrie. *Math. Ann.* **4**, 573–625.

F. Klein [1928], *Nicht-Euklidische Geometrie*, Springer, Berlin.

J. Liouville [1850], Note IV to Monge's *Application de l'analyse à la geometrie*, 5th edition. Bachelier, Paris.

J. Milnor [1982], Hyperbolic geometry: The first 150 years. *Bull Amer. Math. Soc.* (new series) **6**, 9–24.

F. Minding [1838], Über die Biegung krummer Flächen. *J. reine angew. Math.* **18**, 365–368.

F. Minding [1840], Beiträge zur Theorie der kürzesten Linien auf krummen Flächen. *J. reine angew. Math.* **20**, 323–327.

H. Poincaré [1985], *Papers on Fuchsian Functions*, Springer-Verlag.

D.J. Struik [1961], *Lectures on Classical Differential Geometry*, Addison-Wesley.

Essay on the Interpretation of Noneuclidean Geometry

by Eugenio Beltrami

***Giornale di Matematiche* 6 (1868) 284–312**

In recent times the mathematical public has begun to take an interest in some new concepts which seem destined, if they prevail, to change profoundly the whole complexion of classical geometry.

These concepts are not particularly recent. The master GAUSS grasped them at the beginning of his scientific career, and although his writings do not contain an explicit exposition, his letters confirm that he had always cultivated them and attest his full support for the doctrine of LOBACHEVSKY.

Such attempts at radical innovation in basic principles are encountered not infrequently in the history of ideas. Today they are a natural result of the critical spirit which accompanies all scientific investigation. When these attempts are presented as the fruits of conscientious and sincere investigations, and when they receive the support of a powerful, undisputed authority, it is the duty of men of science to discuss them calmly, avoiding equally both enthusiasm and disapproval. Moreover, in the science of mathematics, the triumph of new concepts cannot negate the truth already gained: it can only change the context or interpretation of the reasoning, and increase or diminish its value and frequency of use. A critique of principles cannot damage the solidity of the scientific edifice, even when it does not lead to the discovery and better understanding of its true foundations.

In this spirit we have sought, to the extent of our ability, to convince ourselves of the results of LOBACHEVSKY'S doctrine; then, following the tradition of scientific research, we have tried to find a real substrate for this doctrine, rather than admit the necessity for a new order of entities and concepts. We believe we have attained this goal for the planar part of the doctrine, but we believe that it is impossible to proceed further.

The present work is intended primarily to develop the first of these theses; the second is simply summarised briefly at the end, in order to allow the most straightforward judgement of the inherent significance of the proposed interpretation.

To avoid frequent interruptions to the exposition, we have postponed some necessary analytic results until a special note at the end.

* * * * *

The fundamental principle of proofs in elementary geometry is the *superimposability of equal figures.*

This principle is applicable not only to the plane, but to all surfaces on which there are equal figures in different positions, that is to say, to all surfaces of which any portion can be mapped onto any other by simple flexion. One sees that the rigidity of the surface on which the figure lies is not an essential condition for the application of the principle, so that, e.g., it does not affect the accuracy of proofs in euclidean plane geometry if the figure lies on the surface of a cylinder or cone, rather than on a plane.

The surfaces with the above property, by a celebrated theorem of GAUSS, are those which have a constant product of principal curvatures at all points, so that the spherical curvature[1] is constant. The surfaces which do not satisfy the principle of superposition for figures traced on them therefore have a structure which varies with position.

The most essential figure in elementary geometry is the straight line. Its specific character is that of being completely determined by two points, so that two lines which pass through the same points coincide throughout their extension. But in plane geometry this characteristic is not used exhaustively, since, as a matter of fact, the line is not needed in considerations of planimetry, thanks to the following postulate: when fitting together two planes, each containing a line, it suffices to superimpose the lines at two points, because then they coincide over their full extension.

Now this characteristic, circumscribed in this way, is not peculiar to straight lines in the plane; it also holds (in general) for geodesics on a surface of constant curvature. A geodesic on any surface already has the property (generally speaking) of being determined by two of its points. But the surfaces of constant curvature, and only these, have the property analogous to that of the plane, namely: given two surfaces of constant and equal curvatures, in each of which there is a geodesic, then superposition of the two surfaces at two points of the geodesic causes them to coincide (in general) along its whole extension.

It follows that, except in the case where this property is subject to exceptions, the theorems of planimetry proved by means of the principle of

[1]Translator's note. Gaussian curvature, or simply curvature.

superposition and the straight line postulate, for plane rectilinear figures, also hold for figures formed analogously on a surface of constant curvature by means of geodesics.

In this way one finds many analogies between the geometries of the sphere and the plane – where the straight lines correspond to geodesics, i.e. great circles – analogies which have been noted in geometry for a long time. If other analogies, of different type but the same origin, have not been given equal attention, it is probably because the idea of mapping flexible surfaces onto each other has not become familiar until recently.

We have already alluded to exceptions which may interrupt or limit the analogy we are discussing. These exceptions really exist. On the spherical surface, for example, two points fail to unambiguously determine a great circle when they are diametrically opposite. This is the reason why certain theorems of planimetry lack full analogues on the sphere, e.g. the following: two perpendiculars to the same line do not meet.

These reflections are the starting point of our present research. We have begun by noting that the conclusion of a proof necessarily embraces all situations in which the hypotheses of the proof are satisfied. If the proof is stated in terms of a particular category of entities, without actually using any properties which differentiate them from a more extensive category, then it is clear that the conclusion of the proof acquires a generality greater than that originally sought. It may very well happen that there are consequences seemingly incompatible with the nature of the entities originally contemplated, inasmuch as a property which holds generally for a given category of entities may be modified or disappear entirely for some particular ones. The apparent incongruity of the results of this kind of investigation, which the mind cannot reconcile, is due to an initially inadequate consciousness of the generality of the investigation.

This being understood at the outset, we consider which proofs in planimetry depend only on the principle of superposition and the postulate of the line, which are exactly those of noneuclidean planimetry. The conclusions of such proofs hold unconditionally wherever this principle and postulate are satisfied. Such situations are necessarily covered by the doctrine of surfaces of constant curvature, but they may not extend to surfaces with exceptional points. The principle of superposition in fact does not suffer from exceptions, but we have seen that the line postulate (for geodesics) meets with exceptions on the sphere, and consequently on all surfaces of positive curvature. Are there also exceptions on surfaces of constant negative curvature? That is to say, can there be two points on such a surface which do not determine a unique geodesic?

This question is apparently still open. If it could be proved that such exceptions are impossible, it would become evident *a priori* that the theorems of noneuclidean planimetry hold unconditionally on all surfaces of constant negative curvature. In that case, certain results which seem incompatible with the properties of the plane become interpretable on such surfaces, and receive a completely satisfactory explanation. At the same time we can explain the passage from euclidean to noneuclidean planimetry in terms of the difference between the surfaces of zero curvature and the surfaces of constant negative curvature.

Such are the considerations which have guided the following research.

* * * * *

The formula

$$ds^2 = R^2 \frac{(a^2 - v^2)du^2 + 2uvdudv + (a^2 - u^2)dv^2}{(a^2 - u^2 - v^2)} \tag{1}$$

represents the square of the line element on a surface whose spherical curvature is constant, negative, and equal to $-\frac{1}{R^2}$. The form of this expression, although less simple than that of other equivalent expressions obtainable by change of variables, has the particular advantage (from our point of view) that a linear equation in u, v represents a geodesic and, conversely, any geodesic is representable by a linear equation in these variables (see Note I at the end).

In particular, the two systems of coordinate lines $u =$ const., $v =$ const. consist of geodesics, whose mutual positions are easily discerned. In fact, if we let θ denote the angle between the two coordinate curves at the point (u, v) we have

$$\cos\theta = \frac{uv}{\sqrt{(a^2 - u^2)(a^2 - v^2)}}, \qquad \sin\theta = \frac{a\sqrt{a^2 - u^2 - v^2}}{\sqrt{(a^2 - u^2)(a^2 - v^2)}} \tag{2}$$

so for $u = 0$ or $v = 0$ we have $\theta = 90°$. Thus the geodesic components of the system $u =$ const. are all orthogonal to the geodesic $v = 0$ of the other system, and the geodesics of the system $v =$ const. are all orthogonal to the geodesic $u = 0$ of the first system. In other words, the point $(u = v = 0)$ is the intersection of the orthogonal geodesics $u = 0$, $v = 0$ which we take as *fundamental*, and any point of the surface is determined as the intersection of geodesics perpendicular to the fundamental ones; this is an obvious generalisation of the ordinary cartesian method.

The formula (2) admits values of the variables u, v subject to

$$u^2 + v^2 \leq a^2. \tag{3}$$

Within these limits the functions E, F, G are real, single-valued, continuous and finite, and E, G, $EG - F^2$ are also positive and nonzero. Then, as we have shown from first principles in the memoir *Sulle variabili complesse in una superficie qualunque* (*Annali di Matematica*, Series II, vol. I), the portion of the surface bounded by the curve

$$u^2 + v^2 = a^2 \tag{4}$$

is simply connected, and the net of geodesic coordinate curves has, at every point, the character of perpendicular parallel systems in the plane, i.e. two geodesics of the same system have no point in common, and two geodesics of different systems intersect. It follows that, in the region considered, each pair of real values u, v satisfying (3) corresponds to a unique real point and, conversely, each point corresponds to a single pair, u, v of real values satisfying the above condition.

Thus, if x, y denote the rectangular coordinates of an auxiliary plane, the equations

$$x = u, \quad y = v$$

establish a map of the region in question, a map for which each point of the region corresponds to a unique point of the plane and conversely. The whole region in question is mapped onto the inside of a circle of radius a with centre at the origin, the boundary being called the *limit circle.* In this map the geodesics of the surface are represented by chords of the limit circle, and in particular the coordinate geodesics are represented by chords parallel to the coordinate axes.

We now see how this limits the region on the surface to which the preceding considerations apply.

A geodesic issuing from the point $(u = 0, v = 0)$ can be represented by the equations

$$u = r\cos\mu, \quad v = r\sin\mu, \tag{5}$$

where r and μ are the ordinary polar coordinates of the point (u, v) on the straight line (on the auxiliary plane) representing the geodesic in question. Then, since μ is constant, it follows from (1) that

$$d\rho = R\frac{adr}{a^2 - r^2}, \quad \text{whence} \quad \rho = \frac{R}{2}\log\frac{a+r}{a-r},$$

where ρ is the arc length of the geodesic containing the point $(u = v = 0)$. We can also write

$$\rho = \frac{R}{2}\log\frac{a + \sqrt{u^2+v^2}}{a - \sqrt{u^2+v^2}} \tag{6}$$

where u, v are the coordinates of the other end of the geodesic arc of length ρ. The square root $\sqrt{u^2+v^2}$ must be taken as positive in order to obtain a positive value for the distance ρ.

This value is zero for $r = 0$ and it grows indefinitely as r or $\sqrt{u^2+v^2}$ increases from 0 to a, becoming infinite for $r = a$ and hence for all values of u, v satisfying (4), and imaginary when $r > a$. Thus it is clear that the curve defined by equation (4), representing the limit circle in the euclidean plane, is none other than the locus of infinite points of the surface, a locus which can be regarded as a geodesic circle with centre at the point $(u = v = 0)$ and infinite (geodesic) radius. This geodesic circle of infinite radius does not exist except in the imaginary or ideal region of the surface, since the region considered a moment ago extends indefinitely far in all directions to embrace all real points of the surface. In this way the limit circle contains a representation of the whole real part of our surface, since all lines are determined by their points at infinity on the limit circle, just as they are determined by their points on concentric internal circles, which are geodesic circles on the surface itself with centre at point $(u = v = 0)$.

If r is taken as a constant in the equation (5), and μ as a variable, then the equations represent a geodesic circle, and formula (1) gives

$$\sigma = \frac{Rr\mu}{\sqrt{a^2-r^2}}, \tag{7}$$

where σ is the arc length on the geodesic circle, represented in the auxiliary plane by the circular arc of radius r and angle μ. Since σ is proportional to μ for any r, we easily see that the ρ geodesics, at their common origin, make the same angles as their counterparts in the auxiliary plane; and that the infinitesimal part of the surface around the point $(u = v = 0)$ is similar to its image on the plane, a property which does not hold for any other point.

It follows from (6) that

$$r = \sqrt{u^2+v^2} = a \tanh\frac{\rho}{R}, \quad \text{and} \quad \cosh\frac{\rho}{R} = \frac{a}{w}, \tag{7$'$}$$

where w denotes the positive value of $\sqrt{a^2-u^2-v^2}$. By virtue of this expression for r, (7) can be written

$$\sigma = \mu R \sinh\frac{\rho}{R},$$

as a result of which the semiperimeter of a geodesic circle of radius ρ is given by

$$\pi R \sinh\frac{\rho}{R} \quad \text{or} \quad \frac{1}{2}\pi R\left(e^{\frac{\rho}{R}} - e^{-\frac{\rho}{R}}\right). \tag{8}$$

It follows from the above that the geodesics on the surface are represented in their (real) totality by chords of the limit circle, while their prolongations past the circle are devoid of a real interpretation. On the other hand, two real points of the surface are represented by two, likewise real, points inside the limit circle, which determine *one* chord of the circle itself. We therefore see that two real points of the surface, *however chosen*, are always connected by a *unique geodesic*, which is represented on the auxiliary plane by the chord through the corresponding two points.

Thus the surfaces of constant negative curvature do *not* admit the exceptions which occur in the case of constant positive curvature, and the theorems of noneuclidean planimetry apply to them fully. These theorems, instead of lacking a concrete interpretation, refer rather to such surfaces, as we now proceed to demonstrate at length. To avoid circumlocution, we call the surface of constant negative curvature *pseudospherical*, and we retain the term *radius* for the constant R on which its curvature depends.

* * * * *

We first seek the general relation between angles of geodesics and the angles of the chords representing them.

Let (u, v) be a point of the surface and let (U, V) be any point on a geodesic issuing from it. Let the equations of two such geodesics be

$$V - v = m(U - u), \quad V - v = n(U - u).$$

Letting α denote the angle between the geodesics at the point (u, v), we have the known formula

$$\tan\alpha = \frac{(n-m)\sqrt{EG - F^2}}{E + (n+m)F + mnG}$$

i.e., for the actual values of E, F, G

$$\tan\alpha = \frac{a(n-m)w}{(1+mn)a^2 - (v - mu)(v - nu)}.$$

Letting α' denote the angle between the two chords, and letting μ, ν be the angles between the latter and the x axis, we have $m = \tan\mu$, $n = \tan\nu$, $\alpha' = \nu - \mu$, and hence

$$\tan\alpha = \frac{aw\sin\alpha'}{a^2\cos\alpha' - (v\cos\mu - u\sin\mu)(v\cos\nu - \sin\nu)}.$$

The denominator of the second expression remains finite at all real points of the surface, hence the angle α cannot be zero unless the numerator is zero. But $\sin\alpha'$ is nonzero as long as the two chords do not coincide. Thus α is nonzero except when $w = 0$, i.e. when the point of intersection of the two geodesics is at infinity.

As a result, we can formulate the following rules:

I. Two distinct chords which intersect inside the limit circle correspond to geodesics which intersect at a point of finite distance and at angle different from 0° or 180°.

II. Two distinct chords which intersect on the limit circle periphery correspond to two geodesics which converge to a single point at infinity and which are inclined at a zero angle to each other.

III. Finally, two distinct chords which intersect beyond the limit circle, or which are parallel, correspond to two geodesics with no common point in their full extensions on the (real) surface.

Now let pq be any chord of the limit circle, and let r be a point inside the circle but not on the chord.

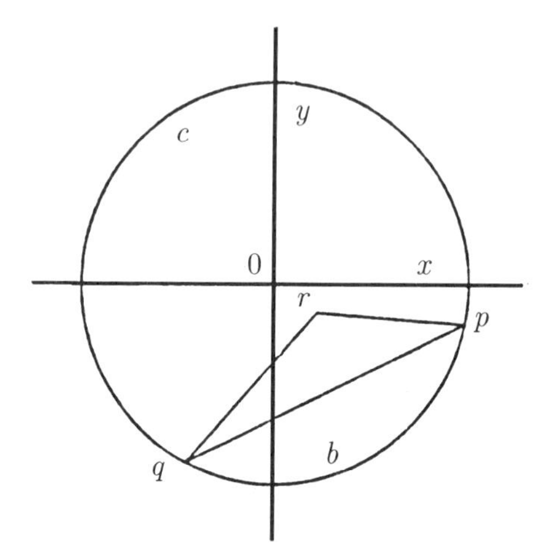

This chord corresponds to a geodesic on the surface, $p'q'$, connecting the points p', q' at infinity (which correspond to p, q); the point r corresponds to a point r' at a finite distance from the geodesic $p'q'$. From this point extend infinitely many geodesics, some of which meet $p'q'$, and some of which do not. The former are represented by the lines from the point r to the arc pbq ($< 180°$), the latter by the lines from this point to the points of the arc pcq ($> 180°$). Two special geodesics divide one of these classes from the other: they are represented by the lines rp, rq, namely, the geodesics issuing from r' which meet $p'q'$ at infinity, one on one side, the other on the other. Since the rectilinear angles rqp, rpq have their vertices on the limit circle, it follows from (II) that the corresponding geodesic angles are zero, although at first sight they appear finite. On the other hand, since r is inside the limit circle and not on the chord pq, the angle prq is different from 0° or 180°, so it follows from (I) that the corresponding geodesics $r'p', r'q'$ make an angle at r' different from 0° or 180°. Thus the geodesics $r'p', r'q'$ can be called *parallels* to $p'q'$ in the sense that they separate the lines which intersect $p'q'$ from those which do not, and we can state the following result: through any (real) point of the surface there are always *two* (real) geodesic parallels to a given (real) geodesic not passing through that point, and which meet each other at an angle different from 0° or 180°.

This result agrees, except in the manner of its expression, with that which forms the pivot of noneuclidean geometry. To see at once, in pseudospherical geometry, the interpretation of noneuclidean geometry, we consider a geodesic triangle. Everyone who studies figures on a surface which is not developable onto the plane has frequent recourse to drawings of figures in the plane which assist the intuition without being strictly geometrically determined; they nevertheless serve to *indicate* the general situation approximately, reproducing the most important positional relations. For a figure to do this it is necessary that the measures of lines and angles be replaced by others of the same type; it is also necessary that the lengths of corresponding lines, and the magnitudes of corresponding angles, stand in a finite ratio to each other, though it does not matter if this ratio varies from place to place in the figure, as long as it does not become zero or infinite. It is obvious that there is a lot of latitude even when the aforementioned ratios do not deviate excessively from some mean value. Under these conditions, if the geodesic triangle just considered has all its vertices at a finite distance, then it is clear that any plane triangle will serve to represent it. This plane triangle can be precisely the rectilinear plane triangle which is its representative in the auxiliary plane, a triangle which lies completely inside the limit circle. Again (and preferably, in some circumstances) it could be a curvilinear triangle whose angles equal those of the geodesic triangle. But if we suppose that the vertices of the geodesic triangle move away into the infinite distance, it is clear that while the triangle continues to be a figure on the surface, with all its points except the vertices at a finite distance, the figure representing it cannot be finite in every sense without violating in some way the conditions we have formulated. For example, a rectilinear triangle representing the geodesic triangle on the auxiliary plane has finite angles, while those of the geodesic triangle are zero. In a curvilinear triangle whose sides are tangential at the vertices, points b, c on the sides which meet at vertex a determine intervals ab, bc whose lengths are in a finite ratio, violating the condition of infinite ratio in the geodesic triangle (Fig. 1a).

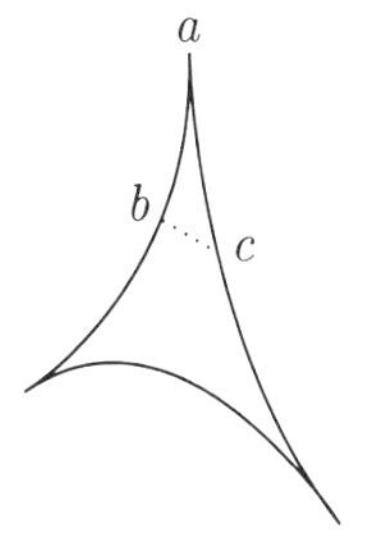

Figure 1a

To remove this discord, it is necessary that all intervals analogous to bc be zero, and this cannot be achieved except by the disposition of Fig. 2a,

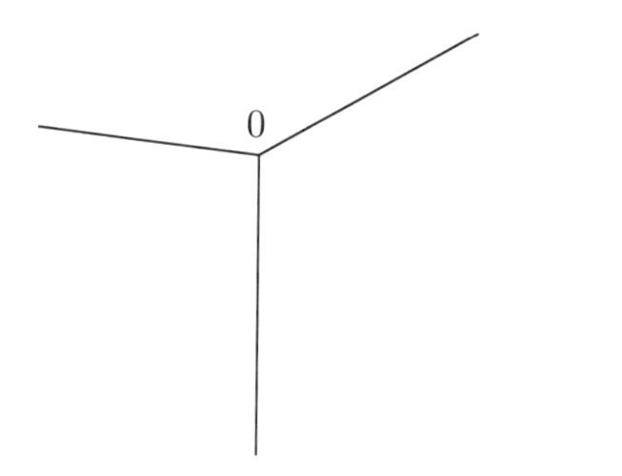

Figure 2a

in which the representatives of all points at a finite distance on the geodesic triangle are concentrated at 0. Such a figure can be conceived as the result of observing the geodesic triangle through a lens endowed with the (fictitious) property of infinite reduction. Under such a hypothesis all the finite intervals appear as zero, and the infinite intervals as finite.

This agrees substantially with what GAUSS noted in his letter of 12 July 1831 to SCHUMACHER,[2] in which he also states that the semiperimeter of the noneuclidean circle of radius ρ has the value

$$\frac{1}{2}\pi k\left(e^{\frac{\rho}{k}}-e^{-\frac{\rho}{k}}\right),$$

where k is a constant. This constant, which GAUSS said may be detectable by measurement of extremely large distances, is none other, according to the theory behind formula (8), than the radius of the pseudospherical surface which we have introduced unconsciously into planimetry in place of the euclidean plane. All our considerations are assumed to apply as much to surfaces of this class as to the plane.

We now proceed to show more concretely how pseudospherical geometry agrees with noneuclidean planimetry. To do this it will be necessary to examine closely the analytic expression we have used to represent the line element of the pseudospherical surface. The following question presents itself initially: must the two fundamental geodesics be chosen in any particular way in order that the line element have its aforementioned form? It certainly seems that they could be chosen arbitrarily, since all pieces of the surface are superimposable on the surface itself, and it is clear then that any two orthogonal geodesics existing in one part are superimposable on any two others, provided the latter are also orthogonal. However, since this question is essential for our purpose, we have treated it in Note II, in which, by

[2]See the appendix to HOÜEL's translation of LOBACHEVSKY's *Theory of Parallels.*

showing directly that the two fundamental geodesics are arbitrary, we have proved at the same time that it is not necessary to admit the preliminary assumption that all parts of the surface be superimposable.

In consequence of this fact and the reasons already stated, the theorems of noneuclidean planimetry for rectilinear figures necessarily become valid for the analogous geodesic figures on a pseudospherical surface. Examples of such are nos. 3-10, 16-24, 29-30 of LOBACHEVSKY'S *Theory of Parallels.*

We now consider the two geodesics from a given point that are parallel to a given geodesic. Let δ be the length of the geodesic normal from the point to the geodesic in question. This normal halves the angle between the parallels. In fact, if we detach the strip of surface bounded by the normal, one parallel, and the corresponding half of the geodesic, and reapply it to the surface so that the normal coincides with its original position, while the geodesic is applied to its other half, it is clear that the parallel bounding the strip becomes applied to the other parallel, otherwise there would be more than two parallels to the given geodesic. The angle between either parallel and the normal is called the *angle of parallelism* and denoted by Δ. To calculate this angle, we make use of our usual analysis, putting the origin $(u = v = 0)$ at the given point and drawing the fundamental geodesic $v = 0$ normal to the given geodesic. The latter then becomes represented by the equation

$$u = a \tan \frac{\delta}{R},$$

as follows easily from the formula (7′).

On the auxiliary plane, the counterpart of this geodesic is a chord perpendicular to the x axis and bisected by it, one end of which has ordinate $\frac{a}{\cosh \frac{\delta}{R}}$. This point on the limit circle determines the radius with equation

$$y = \frac{x}{\sinh \frac{\delta}{R}},$$

which corresponds on the surface to one of the parallels considered; and since angles at the origin are the same on the surface and on the auxiliary plane, we evidently have

$$\tan \Delta \cdot \sinh \frac{\delta}{R} = 1, \tag{9}$$

a formula which expresses the relation between the length δ of the normal and the angle of parallelism Δ. This agrees with that found by BATTAGLINI (see this journal, p. 225). To compare with LOBACHEVSKY, we rewrite it in the form

$$e^{-\frac{2\delta}{R}} + 2e^{-\frac{\delta}{R}} \cot \Delta - 1 = 0$$

and deduce

$$e^{-\frac{\delta}{R}} = \frac{-\cos\Delta \pm 1}{\sin\Delta}.$$

The lower sign is inadmissible because $\frac{\delta}{R}$ is a real quantity, hence

$$\tan\frac{\Delta}{2} = e^{-\frac{\delta}{R}},$$

which is exactly the formula of LOBACHEVSKY (*Theory of Parallels*, no. 36), except for a difference in symbols due to choice of the unit.

Denoting the angle of parallelism relative to a normal distance z by $\Pi(z)$, as does LOBACHEVSKY (no. 16), we obtain from (9) that

$$\cosh\frac{z}{R} = \frac{1}{\sin\Pi(z)}, \qquad \sinh\frac{z}{R} = \cot\Pi(z). \tag{10}$$

Now, by an observation of MINDING (vol. XX of CRELLE'S journal), developed by CODAZZI (in the *Annali* of TORTOLINI, 1857), the ordinary formulae for spherical triangles are converted into those for geodesic triangles on a surface of constant negative curvature by inserting the factor $\sqrt{-1}$ in the ratio of the side to radius and leaving the angles unaltered, which amounts to changing the circular functions involving the radius into hyperbolic functions. For example, the first formula of spherical trigonometry

$$\cos\frac{a}{R} = \cos\frac{b}{R}\cos\frac{c}{R} + \sin\frac{b}{R}\sin\frac{c}{R}\cos A$$

becomes

$$\cosh\frac{a}{R} = \cosh\frac{b}{R}\cosh\frac{c}{R} + \sinh\frac{b}{R}\sinh\frac{c}{R}\cos A.$$

Replacing the letters a, b, c by the corresponding angles of parallelism by means of the formula (10), this relation is converted into the following:

$$\cos A\cdot\cos\Pi(b)\cdot\cos\Pi(c) + \frac{\sin\Pi(b)\cdot\sin\Pi(c)}{\sin\Pi(a)} = 1,$$

which is one of the fundamental equations of noneuclidean planimetry (*Theory of Parallels*, no. 37). The others can be obtained analogously. (The inverse passage from these equations to those of spherical trigonometry was indicated by LOBACHEVSKY, p. 45, but simply as an analytic fact.)

The preceding results fully exhibit the correspondence between noneuclidean planimetry and pseudospherical geometry. To verify the same fact from another point of view, we shall establish directly, by our analysis, the theorem about the sum of the three angles of a triangle.

Consider a right triangle formed by the fundamental geodesic $v = 0$, a perpendicular geodesic $u =$ const., and the geodesic issuing from the origin at angle μ, whose equation is

$$v = u\tan\mu.$$

We let μ' be the third angle of this triangle. The corresponding angle in the auxiliary plane is $90° - \mu$, hence, by the relation previously established between corresponding angles on surface and plane we have

$$\tan\mu' = \frac{w\cos\mu}{a\sin\mu},$$

from which it can be seen that, since μ is an acute angle, so is μ'. Since $v = u\tan\mu$, this formula can be written, using the positive square root, as

$$\tan\mu' = \frac{\sqrt{a^2\cos^2\mu - u^2}}{a\sin\mu}, \quad \text{whence} \quad d\mu' = \frac{a\sin\mu\cdot udu}{(a^2-u^2)\sqrt{a^2\cos^2\mu - u^2}},$$

so that μ' decreases when μ is held constant and the side opposite it is moved away. We take the element of surface area

$$dudv\sqrt{EG-F^2} = R^2 a\frac{dudv}{(a^2-u^2-v^2)^{3/2}}$$

and integrate with respect to v from $v = 0$ to $v = u\tan\mu$, obtaining

$$\frac{R^2 a\sin\mu\cdot udu}{(a^2-u^2)\sqrt{a^2\cos^2\mu - u^2}}, \quad \text{that is,} \quad -R^2 d\mu',$$

for the increment of area in the triangle considered when the side opposite the angle μ is moved. Integrating again from $\mu' = 90° - \mu$ to $\mu' = \mu$ (the first of which evidently corresponds to $u = 0$) we obtain

$$R^2(\frac{\pi}{2} - \mu - \mu')$$

as the area of the right triangle. By dividing an arbitrary geodesic triangle ABC into right triangles by a geodesic normal from one vertex to the opposite side, one finds its area to be

$$R^2(\pi - A - B - C).$$

This expression, since it must be positive, shows that the sum of the three angles of any geodesic triangle cannot exceed 180°. For it to equal 180° in any finite triangle, it is necessary that $R = \infty$, and then $A + B + C = \pi$ in any other finite triangle. But when $R = \infty$, (9) gives $\Delta = \frac{\pi}{2}$, whence the angle of parallelism is necessarily right; and conversely. These are also the conclusions of noneuclidean geometry.

The triangle formed by a geodesic and its two geodesic parallels through an external point has two angles zero and the third equal to 2Δ, hence its area is finite and equal to

$$R^2(\pi - 2\Delta) \quad \text{which, by (9), is} \quad 2R^2\tan^{-1}(\sinh\frac{\delta}{R}),$$

where δ is the distance from the point to the geodesic. When R is very large this quantity approximately equals $2\delta R$, and it is therefore infinite for the plane but not, as we have seen, in the present case.

A geodesic triangle whose vertices are all at infinity has a finite area, namely πR^2, which is independent of its form.[3]

A geodesic polygon with n sides, and internal angles A, B, C, ..., has the area

$$R^2\{(n-1)\pi - A - B - C - \cdots\}.$$

If the polygon has all its vertices at infinity then its area, which does not cease to be finite, reduces to $(n-2)\pi R^2$ and hence is independent of its form.

* * * * *

We now examine the curves we have called geodesic circles.

At the end of Note II we find that the geodesic circle with centre at any point (u_0, v_0) and geodesic radius ρ is represented by the equation

$$\frac{a^2 - uu_0 - vv_0}{\sqrt{(a^2-u^2-v^2)(a^2-u_0^2-v_0^2)}} = \cosh\frac{\rho}{R}. \tag{11}$$

This general equation will become useful later, but for the moment we can profit from the simplification which results from taking the origin ($u = v = 0$) as the centre of the geodesic circle in question. Giving the expression for the line element (as in Note II) the form

$$ds^2 = R^2\frac{w^2(du^2+dv^2)+(udu+vdv)^2}{w^4}$$

and putting

$$u = r\cos\varphi, \quad v = r\sin\varphi,$$

we immediately deduce the equivalent expression

$$ds^2 = R^2\left[\left(\frac{adr}{a^2-r^2}\right)^2 + \frac{r^2d\varphi^2}{a^2-r^2}\right].$$

But, taking ρ as the geodesic distance from the point (u, v) or (r, φ) to the origin, we know that

$$\frac{adr}{a^2-r^2} = \frac{d\rho}{R}, \quad \frac{r^2}{a^2-r^2} = \sinh^2\frac{\rho}{R},$$

[3]Translator's note. Beltrami means by this the form of the image in the auxiliary (euclidean) plane. All triangles with vertices at infinity are in fact *congruent* in hyperbolic geometry, hence the same is true of n-gons with all their vertices at infinity.

whence

$$ds^2 = d\rho^2 + \left(R \sinh \frac{\rho}{R}\right)^2 d\varphi^2, \tag{12}$$

which is an expression already known for the line element of a pseudospherical surface.

This expression reappears as the canonical form of the line element of a surface of revolution. However, it is necessary to observe that in the actual case the pseudospherical cap around the point ($u = v = 0$) cannot be effectively mapped onto a surface of rotation without breaking the continuity by a cut through this point. In fact, the supposed surface of revolution which escapes this condition meets its own axis at the common centre ($\rho = 0$) of all the geodesic circles $\rho =$ const., at which point both curvatures have the same sign, which is impossible since all points of a pseudospherical surface are *hyperbolic*. The same impossibility results when, in trying to avoid the aforementioned cut, the variable φ is taken as the longitude of the variable meridian, and consequently the radius of the parallel corresponding to the meridian arc is $R \sinh \frac{\rho}{R}$. The variation of this radius is therefore $\cosh \frac{\rho}{R} \cdot d\rho$, which is $> d\rho$, and this is absurd, because the variation in question is the projection of $d\rho$ onto the plane containing the parallel.

The expression (12) for the line element, although lacking the advantages inherent in the variables u, v, can sometimes be used because of its simplicity. This is the case, e.g., in determining the tangential curvature of a geodesic circle which, when the radius is ρ, has value $\frac{1}{R \tanh \frac{\rho}{R}}$. This curvature is the same at each point of the circle. The latter property can be seen *a priori* by observing that the piece of surface terminated by a geodesic circle can be applied in any way to the surface itself without its boundary ceasing to be a geodesic circle centred on the point to which the centre of the piece is applied.

The theorem that the geodesic normal bisectors of chords of a geodesic circle meet at its centre is proved in the same way as the corresponding theorem of ordinary planimetry, and we conclude that the construction of the centre of the circle through three points not on the same geodesic is analogous to the ordinary one. This circle is therefore unique.

However, a difficulty arises here. Given three arbitrary points of the surface, it can happen that the geodesic perpendicular bisectors of the lines joining them do not meet at any *real* point of the surface. Consequently, if we confine the term *geodesic circle* to a curve described by the extremity of a geodesic arc as it turns about a *real* point of the surface, it will be necessary to admit that it is not always possible to have a geodesic circle through three points of the surface, *chosen arbitrarily*. Again this agrees, *mutatis mutandis*, with the principles of LOBACHEVSKY (*Theory of Parallels*, no. 29).

Nevertheless, since geodesics of the surface can always be represented by chords of the limit circle, if two chords are such that their prolongations meet at a point outside the circle, then it is permissible to regard the corresponding geodesics as having an *ideal* point in common, and their orthogonal trajectories are analogous to geodesic circles with the ideal point as centre.

We now find the equations of these trajectories directly.

The equation

$$v - v_0 = k(u - u_0)$$

represents the system of geodesics issuing from the point (u_0, v_0), which is real or ideal according as $u_0^2 + v_0^2$ is smaller or greater than a^2. The differential equation of this system is

$$\frac{du}{u - u_0} = \frac{dv}{v - v_0},$$

and consequently, that of the orthogonal system will be

$$[E(u - u_0) + F(v - v_0)]du + [F(u - u_0) + G(v - v_0)]dv = 0,$$

which, for the actual values of E, F, G is

$$d\frac{a^2 - uu_0 - vv_0}{\sqrt{a^2 - u^2 - v^2}} = 0.$$

Then

$$\frac{a^2 - uu_0 - vv_0}{\sqrt{a^2 - u^2 - v^2}} = C \tag{13}$$

is the finite equation for the geodesic circle in the general sense, for a real or ideal centre (u_0, v_0).

When this centre is real, its distance from the curve is constant, by virtue of a well-known theorem. In fact, if we denote this distance by ρ we have, by comparing with equation (11),

$$\cosh\frac{\rho}{R} = \frac{C}{\sqrt{a_0^2 - u_0^2 - v_0^2}}.$$

In this case it is clear that the admissible values for the constant C do not include zero, because the locus corresponding to this value would be represented in the auxiliary plane by a line exterior to the limit circle, and thus would fall into the ideal portion of the surface.

When the centre is ideal, the notion of geodesic radius is inapplicable; but the constant C can take the value zero, since the resulting equation

$$a^2 - uu_0 - vv_0 = 0$$

represents, on the auxiliary plane, a chord of the limit circle which is precisely the polar of the external point (u_0, v_0). This equation defines a real geodesic of the surface. Thus we can conclude that, among the infinitely many geodesic circles with a given ideal centre, there is always one (and only one) real geodesic, so that the geodesic circles with ideal centre can also be defined as the curves parallel (geodesically[4]) to a real geodesic. The latter property has been noted by BATTAGLINI (*l.c.* p. 228) in different language. We see then that whereas, on the sphere, the two ideas of *geodesic circle* and *parallel curve to a geodesic* completely coincide, on the pseudospherical surface they are different, reflecting the reality or ideal nature of the centre.

Since each geodesic circle with ideal centre is equidistant, at all its points, from a certain geodesic, we shall assume that the latter is $v = 0$, which is always permissible, and we shall let ξ be the geodesic distance from the point (u, v) to this line. This distance is measured on one of the geodesics of the system $u = \text{const.}$, and is given by the formula

$$\xi = \frac{R}{2} \log \frac{\sqrt{a^2 - u^2} + v}{\sqrt{a^2 - u^2} - v}.$$

Assuming ξ constant, the resulting equation between u and v represents an arbitrary geodesic circle with centre at the ideal point of concurrence of all geodesic normals to the line $v = 0$.

If we let η denote the arc of the geodesic $v = 0$ between the origin and the normal ξ, its value is given by

$$\eta = \frac{R}{2} \log \frac{a + u}{a - u}.$$

The last two equations yield

$$u = a \tanh \frac{\eta}{R}, \qquad v = \frac{a \tanh \frac{\xi}{R}}{\cosh \frac{\eta}{R}},$$

whence

$$w^2 = a^2 - u^2 - v^2 = \frac{a^2}{\cosh^2 \frac{\xi}{R} \cosh^2 \frac{\eta}{R}}.$$

Substituting for the variables u, v in the expression (1) we now find

$$ds^2 = d\xi^2 + \cosh^2 \frac{\xi}{R} \cdot d\eta^2, \tag{14}$$

an expression which fits a surface of revolution.

[4]Translator's note. I.e. at constant distance.

Letting r_0 denote the radius of the minimum parallel of this surface, which evidently corresponds to $\xi = 0$, and letting r denote the radius of the parallel ξ, we have

$$r = r_0 \cosh \frac{\xi}{R}, \quad \text{whence} \quad \frac{dr}{d\xi} = \frac{r_0}{R} \sinh \frac{\xi}{R}.$$

Then the zone of the pseudospherical surface which can actually be mapped onto the surface of revolution is defined by the condition

$$\left(\frac{r_0}{R} \sinh \frac{\xi}{R} \right)^2 < 1,$$ [5]

i.e. it lies between two geodesic circles equidistant from the geodesic $\xi = 0$ which maps onto the minimum parallel. The width of this zone depends on the radius we assign to the minimum parallel, and increases as the latter decreases. The length of the zone is indefinite, and hence it is wrapped infinitely many times round the surface of revolution. To see why this is necessary one observes that points which map to the same position must still be regarded as distinct, otherwise the theorem that two points determine a unique geodesic would cease to be true. In other words, one should view the surface of revolution as the limit of a helicoid as the step distance converges to zero. The two extreme parallels have radius $\sqrt{R^2 + r_0^2}$, and their planes are tangential to the surface.

As intermediate figures between the geodesic circles with real centre and those with ideal centre one has geodesic circles with centre at infinity. The latter merit study because of their remarkable properties.

The general equation of these circles retains the form (13), because the method used to derive it is valid for all positions of the centre; but if this equation is compared with (11), in which the quantity $\sqrt{a^2 - u_0^2 - v_0^2}$ or w_0 converges to zero as the centre tends to infinity while the second term increases indefinitely, we see that the product $w_0 \cosh \frac{\rho}{R}$ converges to a finite value, which is evidently the same as the limit of $\frac{1}{2} w_0 e^{\frac{\rho}{R}}$. Now, if we replace ρ by $\rho' - \rho$, (11) can be written

$$\frac{a^2 - uu_0 - vv_0}{\sqrt{a^2 - u^2 - v^2}} = \frac{w_0}{2} e^{\frac{\rho'}{R}} \cdot e^{-\frac{\rho}{R}} + \frac{w_0}{2} e^{-\frac{\rho'}{R}} \cdot e^{\frac{\rho}{R}}.$$

Then, by keeping ρ finite and letting ρ' increase indefinitely while w_0 tends to zero, we have in the limit

$$\frac{a^2 - uu_0 - vv_0}{\sqrt{a^2 - u^2 - v^2}} = k e^{-\frac{\rho}{R}}$$

[5] Translator's note. Since, as Beltrami remarks below, $\frac{dr}{d\xi} = 1$ implies that the surface is tangential to the plane of the parallel ξ.

where k is a constant. When the geodesic circles with centre (u_0, v_0) at infinity are represented in this way, the parameter ρ expresses the constant interval between an arbitrary one of these circles and a fixed one, and it increases positively from the latter towards the centre at infinity. Setting $k = a$, the circle $\rho = 0$ becomes the one which passes through the point $(u = v = 0)$.

If we combine the resulting equation

$$\frac{a^2 - uu_0 - vv_0}{\sqrt{a^2 - u^2 - v2}} = ae^{-\frac{\rho}{R}} \tag{15}$$

with the equation

$$\frac{u_0 v - uv_0}{a^2 - uu_0 - vv_0} = \frac{\sigma}{R} \tag{16}$$

and take account of the relation $u_0^2 + v_0^2 = a^2$, we find that the line element (1) assumes the form

$$ds^2 = d\rho^2 + e^{-\frac{2\rho}{R}} d\sigma^2, \tag{17}$$

which again arises from a surface of revolution.

Letting r_0 denote the radius of the parallel $\rho = 0$, where σ is the arc length, and letting r be the radius of the parallel ρ, we have

$$r = r_0 e^{-\frac{\rho}{R}},$$

whence the surface of revolution is real only in the interval determined by the relation $\rho > R \log \frac{r_0}{R}$, so that the circle $\rho = 0$ cannot be a real parallel if $r_0 \leq R$. The maximum parallel for radius R corresponds to the value $\rho = R \log \frac{r_0}{R}$. To determine r_0 conveniently, this parallel can be recovered from any one of the circles in question; for example, by taking $r_0 = R$ we obtain the initial circle $\rho = 0$. The minimum parallel corresponds to $\rho = \infty$ and has radius zero, consequently the surface of revolution asymptotically approaches its axis on one side, while on the other side it is bounded by the maximum parallel, the plane of which it meets tangentially. This surface is wrapped, with infinitely many turns, by the portion of the pseudospherical surface bounded by the line $\rho = 0$, when $r_0 = R$.

The tangential curvature of any parallel is found to be $\frac{1}{R}$, i.e. it is the same for all. Now, the radius of tangential curvature of a parallel curve is nothing but the portion of the tangent to the meridian between the point of contact with the surface (on the parallel considered) and the axis. Thus, on the surface of revolution in question, this portion of the tangent is constant. The meridian curve is the well-known *curve of constant tangent*[6] and the surface generated is the one ordinarily regarded as the typical surface of constant negative curvature[7] (LIOUVILLE in Note IV to MONGE).

[6] Translator's note. Tractrix.

[7] Translator's note. Now called *the* pseudosphere.

The geodesic circles with centre at infinity obviously correspond to the *horocycles* in LOBACHEVSKY'S geometry (*Theory of Parallels*, nos. 31 and 32). We can therefore say that a system of concentric horocycles is transformed, by a suitable flexion of the surface, into the system of parallels on the surface of revolution generated by a curve with constant tangent.

To check how our horocycles correspond to those of LOBACHEVSKY, we observe that the dihedral angle $\frac{\sigma}{R}$ between two meridian planes corresponds, on the parallels ρ_1 and ρ_2, to two arcs s_1, s_2 given by the equations

$$s_1 = \sigma e^{-\frac{\rho_1}{R}}, \quad s_2 = \sigma e^{-\frac{\rho_2}{R}}$$

whence, letting τ be the distance $\rho_2 - \rho_1$, we derive

$$s_2 = s_1 e^{-\frac{\tau}{R}},$$

which coincides with the formula of LOBACHEVSKY (no. 33), apart from the usual difference in the choice of unit.

The expression (17) for the line element is independent of the coordinates (u_0, v_0) of the centre of the horocycle considered; moreover, we have seen that any horocycle in the system can take the place of the maximum. We therefore conclude that any horocycle on the surface can be superimposed on any other.

Two points of the pseudospherical surface are always connected by two horocycles, which are determined by constructing the perpendicular geodesic bisector of the geodesic joining them. Its points at infinity are the centres of the horocycles sought. The arcs of these horocycles between the given points are of equal length, which depends on the geodesic distance between the points. If we call this distance ρ, and let σ be the length of the arcs, then we easily find, using equations (15) and (16) (where ρ has a different meaning) that

$$\sigma = 2R \sinh \frac{\rho}{2R},$$

which presents a singular analogy with the well-known formula connecting the lengths of chord and arc in a circle of radius R. (Cf. BATTAGLINI, *l.c.* 229, and also our note in *Annali di Matematica*, vol. VI, 1865, p. 271.)

* * * * *

The preceding seems to confirm at every point the interpretation of noneuclidean planimetry by means of surfaces of constant negative curvature.

The nature of this interpretation is such that there cannot be an analogous, equally real, interpretation of noneuclidean stereometry. In fact, to obtain the interpretation we have given it is necessary to replace the plane by a surface not reducible to it, i.e. one whose line element cannot be reduced in any way to the form

$$\sqrt{dx^2 + dy^2},$$

which essentially characterizes the plane itself. If we lacked the notion of surfaces not applicable to the plane, then it would be impossible for us to attribute a geometric significance to the construction developed above. But analogy naturally leads us to believe that, if there is an analogous construction for noneuclidean stereometry, then this construction will require the consideration of a space whose line element is *not* reducible to the form

$$\sqrt{dx^2 + dy^2 + dz^2}$$

which essentially characterizes euclidean space. And since, until now, the notion of a space different from the ordinary one is lacking, or at least seeming to transcend the domain of ordinary geometry, it is reasonable to suppose that, even though the analytic considerations which support the above constructions can be extended from two variables to three, the results obtained in the latter case will not admit a representation within ordinary geometry.

This conjecture acquires a degree of probability very close to certainty when we try to extend the above analysis to the case of three variables. In fact, setting

$$\begin{aligned} ds^2 = & \frac{R^2}{(a^2 - t^2 - u^2 - v^2)^2} \\ & \{(a^2 - u^2 - v^2)dt^2 + (a^2 - v^2 - t^2)du^2 + (a^2 - t^2 - u^2)dv^2 \\ & \quad + 2uv\, dudv + 2vt\, dvdt + 2tu\, dtdu\} \end{aligned} \tag{18}$$

as the formula which is suggested *a priori* for three variables t, u, v by formula (1) for two variables u, v, it is easy to check that the analytic deductions from the expression (1) go through essentially unchanged for the new expression, and that the resulting value of ds is effectively that for the line element in a space which provides as complete an interpretation for noneuclidean stereometry, *analytically* speaking, as we have given for the planimetry.

But if we replace the three variables t, u, v by three new ones ρ, ρ_1, ρ_2 by setting

$$t = r\cos\rho_1, \quad u = r\sin\rho_1\cos\rho_2, \quad v = r\sin\rho_1\sin\rho_2,$$

$$\frac{Radr}{a^2 - r^2} = d\rho,$$

we find

$$ds^2 = d\rho^2 + (R\sinh\frac{\rho}{R})^2(d\rho_1^2 + \sin^2\rho_1 \cdot d\rho_2^2),$$

a formula which shows that ρ, ρ_1, ρ_2 are orthogonal curvilinear coordinates in the space considered.

But LAMÉ has shown (*Leçons sur les coordonnées curvilignes*, p. 76, 78) that if the curvilinear coordinates of a space with parameters ρ, ρ_1, ρ_2 define three families of orthogonal surfaces, in which case the square of the distance between infinitesimally close points is represented by an expression of the form

$$ds^2 = H^2 d\rho^2 + H_1^2 d\rho_1^2 + H_2^2 d\rho_2^2,$$

then the three functions H, H_1, H_2 of ρ, ρ_1, ρ_2 which occur in this expression are necessarily subject to two systems of three partial differential equations, typified by

$$\frac{\partial^2 H}{\partial\rho_1 \partial\rho_2} = \frac{1}{H_1}\frac{\partial H}{\partial\rho_1}\frac{\partial H_1}{\partial\rho_2} + \frac{1}{H_2}\frac{\partial H}{\partial\rho_2}\frac{\partial H_2}{\partial\rho_1},$$

$$\frac{\partial}{\partial\rho_1}\left(\frac{1}{H_1}\frac{\partial H_2}{\partial\rho_1}\right) + \frac{\partial}{\partial\rho_2}\left(\frac{1}{H_2}\frac{\partial H_1}{\partial\rho_2}\right) + \frac{1}{H_2}\frac{\partial H_1}{\partial\rho}\frac{\partial H_2}{\partial\rho} = 0.$$

In our case $H = 1$, $H_1 = R\sinh\frac{\rho}{R}$, $H_2 = R\sinh\frac{\rho}{R}\sin\rho_1$, and these values satisfy the first three equations identically, but the latter three are not satisfied unless $R = \infty$. Thus the expression (18) cannot represent the line element of ordinary euclidean space, and the formulae found from this expression cannot be interpreted by figures which occur in ordinary geometry.

To complete the proof that it is impossible to obtain a construction of noneuclidean stereometry within the field of ordinary geometry, it would be necessary to exclude the possibility of arriving at the result by other than an extension of the method followed for planimetry. We do not pretend to have proved that this is out of the question, we only say that it seems to us most improbable.

We have said that the expression (18) serves as a basis for a complete analytic interpretation of noneuclidean stereometry.

This interpretation will be expounded in another memoir.[8] Here we observe only that by taking $t =$ const. in (18) one obtains the expression for the line element of a real surface of constant negative curvature; i.e. a surface which we have seen to satisfy the theorems of noneuclidean planimetry, and which can be considered to exist just as much in ordinary space as in noneuclidean space.

[8] To appear in *Annali di Matematica*[9] where the most general principles of noneuclidean geometry are considered independently of their possible relations with ordinary geometric entities. In the present work we have been interested mainly in offering a concrete counterpart of abstract geometry; however, we do not wish to omit a declaration that the validity of the new order of concepts does not depend on the possibility of such a counterpart.

[9] Translator's note. (1868-69), p.232, entitled *Teoria fondamentale degli spazii di curvatura costante*; see next paper in this volume.

NOTE I

The reduction of the line element of a surface of constant negative curvature to the form we have used in the above investigation is found among the results of our memoir in vol. VII of *Annali di Matematica* (Rome 1866), entitled: *Risoluzione del problema di riportare i punti di una superficie sopra un piano in modo che le linee geodetiche vengano rappresentate da linee rette.*

The principle which enabled us to solve this problem is the following: when the points of a surface are made to correspond, in any way, with those of the plane, one can always take the two independent variables u, v which determine a point of the surface to be the rectangular coordinates x, y of the corresponding point of the plane. That being given, if the correspondence is required to map geodesics of the surface onto lines of the plane, it will be necessary for the 2nd order differential equation of geodesics to have a linear equation in u and v as its complete integral, and consequently this differential equation must reduce simply to the following:

$$du \cdot d^2v - dv \cdot d^2u = 0.$$

But it follows from the general form of the equation in question that this cannot occur unless the functions E, F, G in the line element

$$ds = \sqrt{Edu^2 + 2Fdudv + Gdv^2}$$

satisfy four relations which imply that the line element itself must always be of the form

$$ds = R\,\frac{\sqrt{(a^2+v^2)du^2 - 2uvdudv + (a^2+u^2)dv^2}}{a^2+u^2+v^2}$$

where R and a are arbitrary constants. To recognise the nature of the surface given by this form we calculate the expression for spherical curvature (product of reciprocals of the principal radii of curvature), and we find that the value is $\frac{1}{R^2}$, whence we conclude that the surfaces in question have constant spherical curvature, and hence that such surfaces are the only ones which admit a planar representation under the prescribed conditions.

In the memoir cited above we have assumed real constants R and a, since this hypothesis was natural in the original context. It is then observed that the line element gives a spherical surface of radius R, tangent to the plane of representation at the origin, and that the mapping is by central projection, so that the variables u, v are precisely the rectangular coordinates of the projection of the point to which these variables refer.

However, since the values of the constants R and a are really arbitrary, it is permissible to suppose they are imaginary, if this is convenient. And in fact, if we change the constants to $R\sqrt{-1}$ and $a\sqrt{-1}$ the resulting line element corresponds to a surface of constant negative curvature $-\frac{1}{R^2}$, the geodesics of which continue, like the former ones, to be representable by straight lines on the plane, and hence by linear equations in u and v. This is how we have passed from the formulae of the cited memoir to those of the present work. The only essential difference between the two cases is that in the former the variables u, v can take all real values, while in the latter they are bounded by certain limits, which are easy to assign.

* * * * *

NOTE II

Writing the expression for the line element in the form

$$ds^2 = R^2 \frac{w^2(du^2 + dv^2) + (udu + vdv)^2}{w^4}, \tag{1}$$

we see immediately that, to pass from the primitive fundamental geodesics to two others which are also orthogonal and through the same origin, one need only apply the ordinary formulae for transformation of rectangular coordinates in the plane with fixed origin, namely

$$u = u' \cos\mu - v' \sin\mu$$
$$v = u' \sin\mu + v' \cos\mu,$$

where u', v' are the new coordinates and μ is the angle that the new fundamental $v' = 0$ makes with the original $v = 0$. In fact, these formulae imply

$$u^2 + v^2 = u'^2 + v'^2, \quad du^2 + dv^2 = du'^2 + dv'^2,$$

so that (1) becomes

$$ds^2 = R^2 \frac{w'^2(du'^2 + dv'^2) + (u'du' + v'dv')^2}{w'^4} \tag{1'}$$

and retains its original form. (We see that the geodesics perpendicular to those through the origin are represented by chords of the limit circle perpendicular to the diameters representing the latter geodesics. Conversely, for two geodesics which intersect orthogonally at the point (u, v) to be represented by orthogonal lines on the auxiliary plane, it is necessary that one of the geodesics pass through the origin ($u = v = 0$), as one easily sees from the formula in the text for the transformation of angles. This property is obvious for central projection of the sphere.) The length of a geodesic arc issuing from the origin also has the same form in the second system as in the first, being given by

$$\rho = \frac{R}{2} \log \frac{a + \sqrt{u'^2 + v'^2}}{a - \sqrt{u'^2 + v'^2}}. \tag{2}$$

Now we look at the effect of changing the origin.

To do this, we take any point (u_0, v_0) and suppose that the fundamental $v' = 0$ of the second system passes through it; i.e. we assume that $\cos \mu = \frac{u_0}{r_0}$, $\sin \mu = \frac{v_0}{r_0}$ and hence

$$u = \frac{u_0 u' - v_0 v'}{r_0}, \qquad v = \frac{v_0 u' + u_0 v'}{r_0}, \tag{3}$$

where $r_0 = \sqrt{u_0^2 + v_0^2}$. We then form a *third* system, whose coordinates u'', v'' have one fundamental as the geodesic $v' = 0$ and the other as its normal geodesic through the point (u_0, v_0).

At an arbitrary point (u', v') we construct the geodesic perpendicular to $v' = 0$. We let q be its length, and p its distance from the primitive origin (measured along $v' = 0$). Formula (2) immediately gives

$$p = \frac{R}{2} \log \frac{a + u'}{a - u'}, \tag{4}$$

while $(1)'$ easily shows, taking $du' = 0$, that

$$q = \frac{R}{2} \log \frac{\sqrt{a^2 - u'^2} + v'}{\sqrt{a^2 - u'^2} - v'}. \tag{5}$$

The geodesic distance p_0 between the two origins $(u = v = 0)$, (u_0, v_0) equals

$$p_0 = \frac{R}{2} \log \frac{a + r_0}{a - r_0},$$

so that the geodesic arc on the fundamental $v'' = 0$ of the third system (which is the same as $v' = 0$) between the point (u_0, v_0) and the normal q is given by

$$p - p_0 = \frac{R}{2} \log \frac{(a + u')(a - r_0)}{(a - u')(a + r_0)}. \tag{6}$$

But, letting a_0 be the constant analogous to a in the third system, and observing that in this system the quantities analogous to p, q in the second system are $p - p_0$ and q, it is clear by analogy with (4), (5) that we must have

$$p - p_0 = \frac{R}{2} \log \frac{a_0 + u''}{a_0 - u''}, \quad q = \frac{R}{2} \log \frac{\sqrt{a_0^2 - u''^2} + v''}{\sqrt{a_0^2 - u''^2} - v''}.$$

Equating these expressions with those in (6), (5), we obtain two relations, which yield

$$(7) \qquad u'' = \frac{aa_0(u' - r_0)}{a^2 - r_0 u'}, \quad v'' = \frac{a_0 w_0 v'}{a^2 - r_0 u'}, \quad \left(w_0 = \sqrt{a^2 - r_0^2}\right).$$

Strictly speaking, the constant a_0 remains indeterminate, because we only have equations between the ratios $\frac{u'}{a}, \frac{v'}{a}$ and the ratios $\frac{u''}{a_0}, \frac{v''}{a_0}$. It seems most convenient to determine a_0 by the condition that for $u'' = 0$, i.e. $u' = r_0$, we have $v' = v''$, and we then find

$$a_0 = w_0 = \sqrt{a^2 - u_0^2 - v_0^2}.$$

Substituting this value in the preceding formulae gives

$$u' = \frac{a(a_0 r_0 + au'')}{aa_0 + r_0 u''}, \quad v' = \frac{aa_0 v''}{aa_0 + r_0 u''},$$

and substituting these values in $(1)'$ gives

$$ds^2 = R^2 \frac{(a_0^2 - v''^2)du''^2 + 2u''v''du''dv'' + (a_0^2 - u''^2)dv''^2}{(a_0^2 - u''^2 - v''^2)^2}.$$

Thus translation of the origin does not alter the form of the line element, except to replace a by a_0, which changes nothing fundamental.

Finally, to obtain a *fourth* system completely independent from the first, we replace the two fundamentals $u'' = 0$, $v'' = 0$, by two new orthogonal geodesics with origin (u_0, v_0), by setting

$$u'' = u''' \cos\nu - v''' \sin\nu, \quad v'' = u''' \sin\nu + v''' \cos\nu,$$

and we know that such a substitution will not change the form of the line element. We then see that the form originally admitted for the line element is in no way dependent on a particular choice of fundamental geodesics. On the contrary, the point $(u = v = 0)$ can be any point of the surface, and the fundamental geodesic can be any geodesic through this point.

Taking account of the relations found between the coordinates of successive systems, and making the abbreviations

$$p = \frac{au_0}{a_0 r_0} \cos\nu - \frac{v_0}{r_0} \sin\nu, \qquad p_1 = \frac{au_0}{a_0 r_0} \sin\nu + \frac{v_0}{r_0} \cos\nu,$$

$$q = \frac{av_0}{a_0 r_0} \cos\nu - \frac{u_0}{r_0} \sin\nu, \qquad q_1 = \frac{av_0}{a_0 r_0} \sin\nu - \frac{u_0}{r_0} \cos\nu,$$

$$r = \frac{r_0 \cos\nu}{aa_0}, \qquad r_1 = \frac{r_0 \sin\nu}{aa_0},$$

we find the following final relations between the coordinates u, v and u''', v'''

$$u = \frac{u_0 + pu''' - p_1 v'''}{1 + ru''' - r_1 v'''}, \qquad v = \frac{v_0 + qu''' - q_1 v'''}{1 + ru''' - r_1 v'''}.$$

If we consider u, v and u''', v'' to be rectangular coordinates of two planes, these formulae express a homographic correspondence between these planes, a circumstance which has been discussed in the memoir cited in Note I.

If we compare the original expression for the line element as a function of u, v with the final expression as a function of u''', v''', we find that the two expressions become identical when we set

$$\frac{u}{a} = \pm\frac{u'''}{a_0}, \qquad \frac{v}{a} = \pm\frac{v'''}{a_0}$$

and also when

$$\frac{u}{a} = \pm\frac{v'''}{a_0}, \qquad \frac{v}{a} = \pm\frac{u'''}{a_0},$$

the choice of sign being arbitrary in each formula. This shows that the pseudospherical surface, considered to be flexible and inextensible, can be superimposed on itself in such a way that any one of its points (u_0, v_0) comes to occupy the position of any other point $(u = v = 0)$, while any geodesic issuing from the first point (e.g. $v''' = 0$) comes to coincide with any geodesic issuing from the second (e.g. $v = 0$). Moreover, the ambiguity of signs allows the superimposition of two geodesic angles of the same magnitude at these two points, both directly and inversely. For example, the right angle formed by the geodesics $u''' = 0$, $v''' = 0$ can be applied to that formed by the geodesics $u = 0$, $v = 0$, not only by making $u''' = 0$ coincide with $n = 0$ and $v''' = 0$ with $v = 0$, but also by making $u''' = 0$ coincide with $v = 0$ and $v''' = 0$ with $u = 0$. Thus every portion of the surface can be superimposed, both directly and inversely, on any other part. Consequently, if the portion contains a figure (e.g. a geodesic triangle), then this figure will admit all the displacements, on the surface, that a plane figure admits on the plane, while remaining congruent to itself. Of course, this congruence need not concern more than the length of lines and the magnitude of angles, because the *absolute* curvature of lines does not come into consideration here.[10]

The property we have just demonstrated was known previously, but it seems to us that the above proof has the rigour that the nature of our subject requires. Besides, the theorem of GAUSS shows that if a surface has the property in question, then that surface is necessarily of constant spherical curvature.

[10]The *relative* congruence we speak of would be *absolute* congruence for a being whose geometric concepts did not transcend the two-dimensional field of the surface in question, just as ours do not transcend the three dimensions of ordinary space.

We should not fail to note a useful result which follows easily from some of the preceding formulae. The geodesic circle with centre (u_0, v_0) and radius ρ is represented, in the third system, by the equation

$$u''^2 + v''^2 = a_0^2 \tanh^2 \frac{\rho}{R},$$

as follows from formula (6) of the text. But, since $a_0 = w_0 = \sqrt{a^2 - r_0^2}$, (7) of the present Note implies

$$u''^2 + v''^2 = \left(\frac{a_0}{a^2 - r_0 u'}\right)^2 \{a^2[(u' - r_0)^2 + v'^2] - (r_0 v')^2\},$$

and likewise (3) gives

$$u' = \frac{uu_0 + vv_0}{r_0}, \quad v' = \frac{u_0 v - uv_0}{r_0},$$

whence

$$u' - r_0 = \frac{u_0(u - u_0) + v_0(v - v_0)}{r_0}, \quad v' = \frac{u_0(v - v_0) - v_0(u - u_0)}{r_0},$$

and finally

$$\frac{a^2\{(u - u_0)^2 + (v - v_0)^2\} - (u_0 v - uv_0)^2}{(a^2 - uu_0 - vv_0)^2} = \tanh^2 \frac{\rho}{R}.$$

This equation gives the geodesic distance ρ between any two points (u, v) and (u_0, v_0). When these points are infinitely close, it immediately returns the expression for the line element from which we set out.

When rewritten in terms of cosh rather than tanh, the preceding equation assumes the more elegant form

$$\frac{a^2 - uu_0 - vv_0}{\sqrt{(a^2 - u^2 - v^2)(a^2 - u_0^2 - v_0^2)}} = \cosh \frac{\rho}{R}.$$

Translator's Introduction

Beltrami's *Fundamental theory of spaces of constant curvature*

In his essay on the interpretation of noneuclidean geometry, Beltrami [1868] promised an analytic interpretation which was not restricted to 2 dimensions. What followed was an application of Riemann's n-dimensional differential geometry and a clarification of some of Riemann's results. Riemann's famous essay [1854] mentions spaces of constant curvature and gives a form for their line element; a related constant curvature metric, in two dimensions, had already been found by Liouville [1850], and a third such metric is implicit in Cayley [1859]. Beltrami's decisive contribution was to observe the connection with noneuclidean geometry and to achieve an elegant unification of the three resulting models of hyperbolic geometry. By one of the injustices of nomenclature that are so common in mathematics, the three models – which could appropriately be called the Riemann-Beltrami, Liouville-Beltrami and Cayley-Beltrami models – are usually known as the Poincaré disc model, the Poincaré half-plane model and the Klein disc model.

Beltrami's derivation of these three models from a common source has been nicely summarised and geometrically interpreted by Milnor [1982]. I shall restrict the interpretation to 2 or 3 dimensions to enable visualisation, and I shall expand Milnor's explanation at a few points. Beltrami does not give geometric interpretations, just transformations of coordinates, but in dimensions 2 and 3 the geometry is clear.

The starting point is a *hemisphere model* consisting of the open hemisphere

$$x_1^2 + x_2^2 + x^2 = a^2, \quad x > 0, \tag{1}$$

in 3-dimensional (x_1, x_2, x)-space, with the line element

$$ds = R\,\frac{\sqrt{dx_1^2 + dx_2^2 + dx^2}}{x}. \tag{2}$$

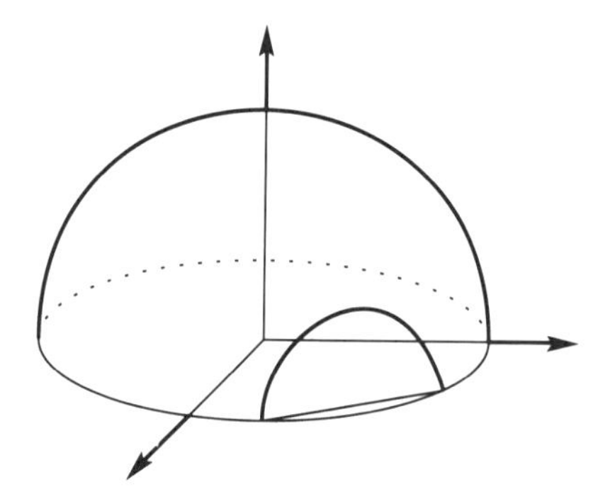

Figure 2.1: Perpendicular projection of hemisphere to disc

One could check that the geodesics for this metric are vertical sections of the hemisphere; however, this also follows from the second model, the "Klein disc," obtained by perpendicular projection onto the plane $x = 0$, which we take to be horizontal (Figure 2.1).

The model is now the open disc $x_1^2 + x_2^2 < a^2$ and the line element is obtained when $x = \sqrt{a^2 - x_1^2 - x_2^2}$ is eliminated from (2) using (1). Beltrami does not do this, presumably because it yields essentially the same expression he had discussed in [1868], but instead proves directly that the geodesics are straight line segments in the disc.

The third model is obtained by stereographic projection of the hemisphere onto its horizontal tangent plane (Figure 2.2), and the line element is computed to be

$$ds = \frac{\sqrt{d\xi_1^2 + d\xi_2^2}}{1 - \frac{1}{4R^2}(\xi_1^2 + \xi_2^2)}, \tag{3}$$

a metric which was stated by Riemann [1854] to be of constant curvature. The transformation to these coordinates is actually called "stereographic" by Beltrami, so he is evidently aware of its geometric interpretation.

We observe that (3) is a multiple of the euclidean line element $\sqrt{d\xi_1^2 + d\xi_2^2}$ in the (ξ_1, ξ_2)-plane. The multiple of course varies with position but not with direction, hence angles are preserved. Thus we have the conformal, or "Poincaré disc" model, 14 years before its appearance in Poincaré [1882].

Presumably Riemann would have also noticed the conformal property of his metric but, since he missed the hyperbolic geometry, he was unable to exploit this geometry in complex analysis as Poincaré did. (See the Poincaré papers in this volume.) Perhaps Beltrami is the missing link between Riemann and Poincaré.

The fourth model is obtained by stereographic projection of the hemisphere onto (the upper half of) a vertical plane (Fig. 2.3). This gives the

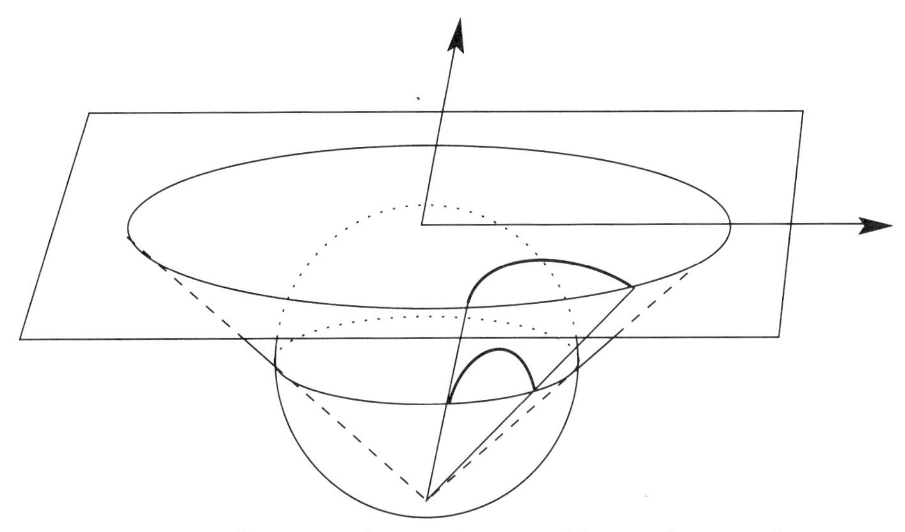

Figure 2.2: Stereographic projection of hemisphere to plane

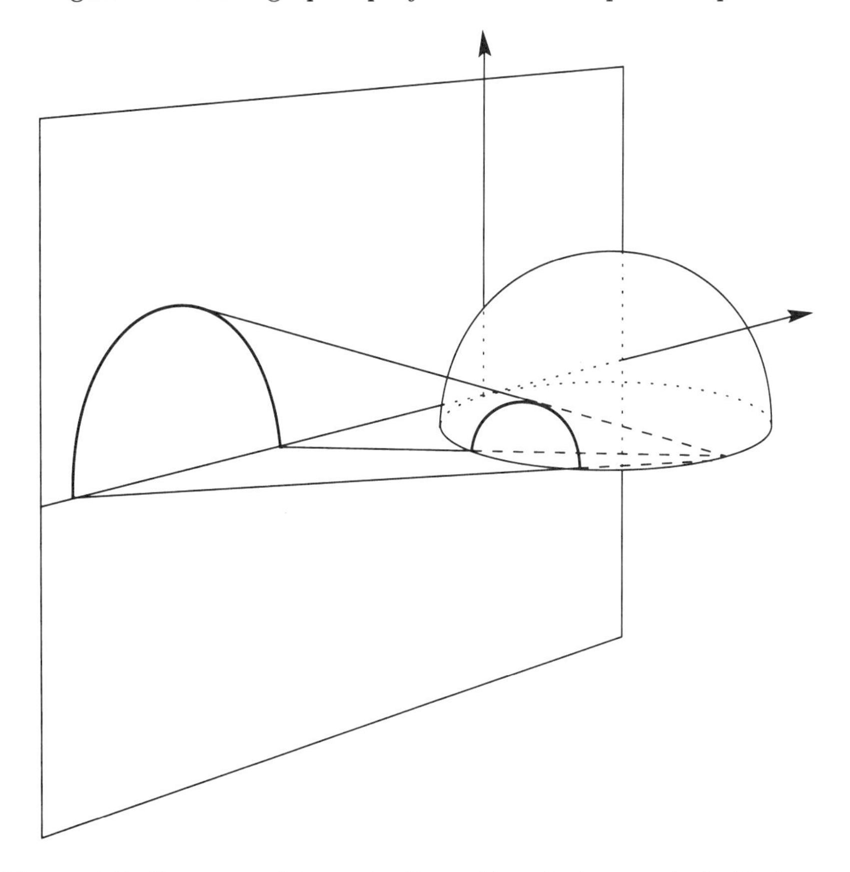

Figure 2.3: Stereographic projection of hemisphere onto half plane

line element

$$ds = r\frac{\sqrt{d\eta_1^2 + d\eta^2}}{\eta}$$

for the "Poincaré half plane," which Liouville [1850] had previously obtained by transformation of the line element for the pseudosphere. This model is also conformal. The n-dimensional generalisation of its line element, for the *half-space model* of n-dimensional hyperbolic geometry,

$$ds = R\frac{\sqrt{d\eta_1^2 + \cdots + d\eta_n^2 + d\eta^2}}{\eta}, \tag{4}$$

is particularly interesting when $n = 3$, because if we write (4) with $x_1 = \eta_1$, $x_2 = \eta_2$, $x = \eta$ we get back the line element (2) we started with. As Beltrami points out, this means that (1) and (2), his hemisphere model, can be viewed as the restriction of the half-space model of hyperbolic 3-space to a hyperbolic plane.

He also points out that, since ds is just the euclidean line element multiplied by a factor which depends only on the height x, the restriction of this metric to a horizontal section $x =$ const. is euclidean. That is, euclidean plane geometry holds on the horizontal sections of the model, which are horospheres in the hyperbolic metric. Thus Beltrami proves with great ease the wonderful result discovered by Wachter in 1816 (see Stäckel [1901]), and proved independently by Bolyai and Lobachevsky, that euclidean geometry is realised within hyperbolic geometry. More generally, Beltrami proves that n-dimensional euclidean geometry is realised within $(n+1)$-dimensional hyperbolic geometry, by the n-dimensional horosphere. (The converse problem, of representing n-dimensional hyperbolic space within some $\mathbb{R}^m$, is in a less satisfactory state. For general n, the smallest known m for which this can be done is $m = 6n - 5$. Gromov and Rokhlin [1970] give a survey of such results.)

At the end of the paper Beltrami mentions the less surprising result that $(n+1)$-dimensional hyperbolic space contains the n-dimensional sphere. Thus hyperbolic geometry can claim to be the "universal" geometry, since it contains the other two constant-curvature geometries in a natural way.

Finally, let us note an interesting connection between the half-space model and the Klein disc which has been pointed out by Thurston [1978]. The vertical lines of (x_1, x_2, x)-space are geodesics in the metric (2), hence if we assume that light rays in hyperbolic space travel along geodesics, an observer can look vertically downwards towards the boundary $x = 0$ (which is at an infinite hyperbolic distance). A hyperbolic plane sufficiently far below is then a hemisphere vertically beneath the observer's eye (Figure 2.4).

The observer's view of this hyperbolic plane is its vertical projection onto $x = 0$, i.e. the Klein disc. Thus the Klein model (which, as we have said,

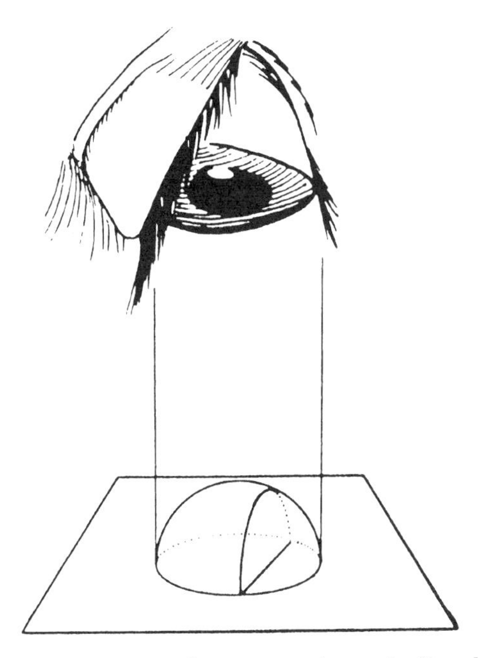

Figure 2.4: Looking down on a hyperbolic plane

is really Beltrami's) represents the *actual view* of a distant hyperbolic plane seen by an observer in hyperbolic space. This suggests that inhabitants of hyperbolic space would have as much trouble visualising hyperbolic geometry as we do! Conceivably, they might prefer to work on horospheres to take advantage of euclidean geometry.

The paper whose translation follows, Beltrami [1868′], was translated into French by Hoüel [1869]. I have consulted Hoüel's translation at points where I found the original Italian difficult.

References

E. Beltrami [1868], Saggio di Interpretazione della Geometria Non-euclidea, *Giornale di Matematiche* VI, 284–312.
[1868′], Teoria fondamentale degli spazii di curvatura costante, *Ann. Mat. pura appl.* ser. II, 232–255.

A. Cayley [1859], A sixth memoir upon quantics. *Mathematical papers*, vol. 2, 561–606.

M.L. Gromov & V.A. Rokhlin [1970], Embeddings and immersions in Riemannian geometry, *Russian Math. Surveys* **25**, no. 5, 1–57.

M.J. Hoüel [1869], French translation of Beltrami [1868′]. *Annales Scientifiques, École Normale Supérieure* **6**, 345–375.

J. Liouville [1859], Note IV to Monge's *Application de l'analyse à la géometrie*, 5th ed. Bachelier, Paris.

J. Milnor [1982], Hyperbolic geometry : The first 150 years. *Bull. Amer. Math. Soc.* (new series) **6**, 9–24.

H. Poincaré [1882], Theorie des groupes fuchsiens. *Acta Math.* **1**, 1–62.

G.F.B. Riemann [1854], Über die Hypothesen, welche der Geometrie zur Grunde liegen, *Collected Works*, Dover, New York, 1953, 272–287. A useful translation and commentary is in M. Spivak: *A Comprehensive Introduction to Differential Geometry*, vol. 2, Publish or Perish, 1970.

P. Stäckel [1901], Friedrich Ludwig Wachter, ein Beitrag zur Geschichte der nicht-euklidischen Geometrie. *Math. Ann.* **54**, 49–85.

W.P. Thurston [1978], The geometry and topology of 3-manifolds. *Lecture Notes*, Princeton University Mathematics Dept.

Fundamental theory of spaces of constant curvature

by Eugenio Beltrami

***Annali di Matematica pura ed applicata* series II, II (1868) 232–255**

In a memoir inserted in *Annali di Matematica pura ed applicata*[11] I have sought the surfaces given by the property that their geodesics can be represented by linear equations, and I have found that this property holds only for surfaces of constant curvature and for certain special variables which arise naturally from the analysis of the problem.

In the present work I expound much more general results to which I have been led by further development of this concept, coordinated with some of the principles sketched by RIEMANN in his remarkable posthumous work: *Über die Hypothesen welche der Geometrie zu Grunde liegen*, recently published by DEDEKIND in volume XIII of the Göttingen memoirs. I hope that my researches will help make certain parts of that profound work more intelligible.

There are certain locutions which I frequently use for the sake of brevity; they will not, I believe, destroy or obscure the basis of the theory. The attentive reader will easily understand, without further explanation, that he is completely free to attribute to them a purely analytic significance.

The differential expression

$$ds = R\frac{\sqrt{dx^2 + dx_1^2 + dx_2^2 + \cdots + dx_n^2}}{x} \tag{1}$$

where $x, x_1, x_2, \ldots, x_n$ are $n+1$ variables connected by the equation

$$x^2 + x_1^2 + x_2^2 + \cdots + x_n^2 = a^2 \tag{2}$$

[11] *Risoluzione del problema: riportare i punti di una superficie sopra un piano in modo che le linee geodetiche rengano rappresentate da linee rette*, vol. VII (1865), p. 185.

and R and a are constants, can be considered to represent the *line element*, or the distance between two infinitely close points, in a *space of* n *dimensions*, each *point* of which is defined by a system of values of the n *coordinates* $x_1, x_2, \ldots, x_n$. The form of this expression determines the *nature* of this space.

If we make the abbreviation

$$\Omega = \sqrt{dx^2 + dx_1^2 + \cdots + dx_n^2},$$

the *geodesics* of the space in question are the curves which satisfy the equation

$$\delta \int \frac{\Omega}{x} = 0$$

under the condition $x\delta x + x_1\delta x_1 + \cdots + x_n\delta x_n = 0$. By ordinary transformations of the variation of an integral, the first equation can be expanded as follows:

$$\int \left\{ \delta x \left[\frac{\Omega}{x^2} + d\left(\frac{dx}{x\Omega}\right)\right] + \delta x_1 \cdot d\left(\frac{dx_1}{x\Omega}\right) + \cdots + \delta x_n \cdot d\left(\frac{dx_2}{x\Omega}\right)\right\} = 0,$$

and, on account of the relation between the variations $\delta x, \delta x_1, \ldots, \delta x_n$, this equation gives way to:

$$\frac{\Omega}{x^2} + d\left(\frac{dx}{x\Omega}\right) = kx, \quad d\left(\frac{dx_1}{x\Omega}\right) = kx_1, \quad \ldots, \quad d\left(\frac{dx_n}{x\Omega}\right) = kx_n,$$

where k is a factor to be determined. Now, multiplying these equations by $x, x_1, \ldots, x_n$ respectively and taking the sum, we have

$$d\left(\frac{xdx + x_1dx_1 + \cdots + x_ndx_n}{x\Omega}\right) = k(x^2 + x_1^2 + \cdots + x_n^2),$$

whence $k = 0$ by equation (2), and consequently

$$d\left(\frac{dx}{x\Omega}\right) + \frac{\Omega}{x^2} = 0, \tag{3}$$

$$dx_1 = c_1 x\Omega, \quad dx_2 = c_2 x\Omega, \quad \ldots \quad dx_n = c_n x\Omega, \tag{4}$$

where $c_1, c_2, \ldots, c_n$ are constants. The last n equations, squared and summed, give

$$\Omega = -\frac{dx}{\sqrt{1 - c^2x^2}} \tag{5}$$

where

$$c = \sqrt{c_1^2 + c_2^2 + \cdots + c_n^2}.$$

This value of Ω makes an identity of equation (3), and hence it is unnecessary to take it into account, while equations (4), after eliminating x and subsequent integration, give

$$x_1 = b_1 x_n + b'_1, \quad x_2 = b_2 x_n + b'_2, \quad \dots, \quad x_{n-1} = b_{n-1} x_n + b'_{n-1}.$$

Thus the geodesics of the space in question are represented by $n-1$ linear equations in the n coordinates $x_1, x_2, \dots, x_n$, just like those in the plane and in ordinary space when we use cartesian coordinates, and those on surfaces of constant negative curvature when we use the variables u, v of the memoir cited. Among the systems of geodesics, we note in particular those which are obtained by holding all but one variable constant. There is a line of each of these systems passing through each point of space, corresponding to the coordinate axes of $x_1, x_2, \dots, x_n$, for which each of the other coordinates are zero; we call these the x_1 *system*, x_2 *system*, ..., x_n *system*.

To obtain the length of the geodesic arc ρ between two given points, we observe that (5) gives

$$d\rho = R\frac{\Omega}{x} = -\frac{R\,dx}{x\sqrt{1-c^2x^2}}$$

whence

$$cx = \frac{1}{\cosh\frac{\rho-\rho_0}{R}},$$

where ρ_0 is an arbitrary constant and x is $\sqrt{a^2 - x_1^2 - x_2^2 - \cdots - x_n^2}$. Letting $x_1^0, x_2^0, \dots, x_n^0$ denote the coordinates of the point $\rho = 0$, the origin of the arc, and letting x^0 be the corresponding value of the function x, we have

$$cx^0 = \frac{1}{\cosh\frac{\rho_0}{R}} \tag{6}$$

whence, eliminating c,

$$x = \frac{x^0\cosh\frac{\rho_0}{R}}{\cosh\frac{\rho-\rho_0}{R}},$$

which can be put in the form

$$\frac{x^2\sinh^2\frac{\rho}{R}}{\cosh^2\frac{\rho_0}{R}} = 2xx^0\cosh\frac{\rho}{R} - x^2 - (x^0)^2. \tag{7}$$

On the other hand, the preceding equations give

$$x\Omega = \frac{x^2 d\rho}{R} = \frac{1}{c^2}\cdot d\tanh\frac{\rho-\rho_0}{R}$$

hence, by the equations (4),

$$x_1 = a_1 + \frac{c_1}{c^2}\tanh\frac{\rho-\rho_0}{R}, \quad x_2 = a_2 + \frac{c_2}{c^2}\tanh\frac{\rho-\rho_0}{R}, \quad \dots,$$

or, substituting $x_1^0, x_2^0, \dots$ for the constants $a_1, a_2, \dots$,

$$x_1 - x_1^0 = c_1 x x^0 \sinh \frac{\rho}{R}, \quad x_2 - x_2^0 = c_2 x x^0 \sinh \frac{\rho}{R}, \quad \dots$$

whence, squaring and summing,

$$2(a^2 - x_1 x_1^0 - x_2 x_2^0 - \dots - x_n x_n^0) - x^2 - (x^0)^2 = c^2 x^2 (x^0)^2 \sinh^2 \frac{\rho}{R}.$$

This equation, by virtue of (6), (7), finally gives

(8)

$$\cosh \frac{\rho}{R} = \frac{a^2 - x_1 x_1^0 - x_2 x_2^0 - \dots - x_n x_n^0}{\sqrt{(a^2 - x_1^2 - x_2^2 - \dots - x_n^2)(a^2 - (x_1^0)^2 - (x_2^0)^2 - \dots - (x_n^0)^2)}},$$

as the general formula for the length of a geodesic arc in terms of the coordinates of its endpoints.

Assuming that the variables $x, x_1, \dots, x_n$ and the constants R, a are real, the *limit* of the space of n dimensions under consideration is the *space* of $n-1$ dimensions given by the equation

$$x_1^2 + x_2^2 + \dots + x_n^2 = a^2. \tag{9}$$

Inside this limit, i.e. where

$$x_1^2 + x_2^2 + \dots + x_n^2 < a^2$$

the first space is *continuous* and *simply connected.* It follows from (8) that the points of the limit space are all at an infinite distance.[12]

> "In all the real field that we have defined, the value of ds, given by equation (1), remains constantly positive for any system of values of the ratios
>
> $$dx_1 : dx_2 : \dots : dx_n.$$
>
> If we consider a second system of increments $\delta x_1, \delta x_2, \dots, \delta x_n$, and if we put
>
> $$\delta s^2 = R^2 \frac{\delta x^2 + \delta x_1^2 + \dots + \delta x_n^2}{x^2},$$

[12](Editor's note in the *Opere Matematiche* of Beltrami, vol. 1, p. 409.) The following text in quotes is not in the original memoir, but is in the French translation, pp. 351-353. It is reproduced here to comply with the author's desires, expressed in a letter to GENOCCHI, in which BELTRAMI suggests a modification to the memoir corresponding to that introduced in the above-mentioned translation. [Cf. the letter to the Beltrami Commemoration written by G. LORIA in *Bibliotheca Mathematica*, Dritte Folge, II Band, 4 Heft (1901), p. 419, note.]

the expression

$$ds^2\delta s^2 - R^4\frac{(dx\delta x + dx_1\delta x_1 + \cdots + dx_n\delta x_n)^2}{x^4}$$

can never become negative (by virtue of a well-known transformation to which it is susceptible); consequently

$$\frac{R^2(dx\delta x + dx_1\delta x_1 + \cdots + dx_n\delta x_n)}{x^2 ds\delta s}$$

cannot be greater than one. Thus we can always assign a real angle θ for which we have[13]

$$(9) \qquad dx\delta x + dx_1\delta x_1 + \cdots + dx_n\delta x_n = \frac{x^2 ds\delta s}{R^2}\cos\theta.$$

"This has the important consequence that, in calculating from equation (1) the three values of ds which correspond to the following three systems of values of the variables, taken in pairs,

$$\begin{array}{cccc} (x_1, x_2, & \ldots, & x_n) \\ (x_1 + dx_1, x_2 + dx_2, & \ldots, & x_n + dx_n) \\ (x_1 + \delta x_1, x_2 + \delta x_2, & \ldots, & x_n + \delta x_n), \end{array}$$

we find three numbers realisable by the sides of a rectilinear triangle. We let M, M', M'' be the three values in question, and represent ds by MM', δs by MM''. The values of the system M'' can be deduced from those of the system M' by means of the respective increments

$$\delta x_1 - dx_1, \quad \delta x_2 - dx_2, \quad \ldots, \quad \delta x_n - dx_n$$

given by the latter. Consequently, neglecting infinitesimals of second order, we can put

$$\begin{aligned} \overline{M'M''}^2 &= \frac{R^2}{x^2}[(\delta x - dx)^2 + (\delta x_1 - dx_1)^2 + \cdots + (\delta x_n - dx_n)^2] \\ &= ds^2 + \delta s^2 - 2\frac{R^2(dx\delta x + dx_1\delta x_1 + \cdots + dx_n\delta x_n)}{x^2} \end{aligned}$$

or, by (9)

$$\overline{M'M''}^2 = \overline{MM'}^2 + \overline{MM''}^2 - 2\overline{MM'}\cdot\overline{MM''}\cos\theta,$$

[13] Translator's note. I have retained the numbering of equations from Beltrami's *Opere*, even though it means a second number (9) here.

where θ is a real angle. This equation proves the property claimed, and explains how each system of values of the variables can be assimilated into a *point* defined by its coordinates. In the same way, we can consider two line elements ds, δs to be *orthogonal*, when $\theta = \frac{\pi}{2}$ for these elements, i.e., by (9), when the increments corresponding to d, δ satisfy the condition

$$dx\delta x + dx_1\delta x_1 + \cdots + dx_n\delta x_n = 0, \tag{10}$$

which we can call, for convenience of language, the *orthogonality condition*."

Consider, for example, the space $x_1 = 0$ of $n-1$ dimensions and suppose that two line elements issue from one of its points, a ds in the space itself and a δs along the geodesic of the x_1 system passing through the point. In this case we have

$$x_1 = 0, \quad dx_1 = 0, \quad \delta x_2 = \delta x_3 = \cdots = \delta x_n = 0, \quad \delta x = 0,$$

and hence the orthogonality condition is satisfied. I.e., each geodesic of the x_1 system (or, more generally, of the x_r system) is orthogonal to the space $x_1 = 0$ (or more generally $x_r = 0$) at the point where they meet. In particular, the n coordinate axes are orthogonal at the origin. It can be shown with equal ease that the x_r axis is orthogonal to all the spaces $x_r =$ const. The n geodesics of the $x_1, x_2, \ldots, x_n$ systems through an arbitrary point are perpendicular to the $(n-1)$-dimensional spaces $x_1 = 0$, $x_2 = 0, \ldots, x_n = 0$ analogously to those in the plane and in ordinary space when we use rectangular coordinates. If we let $X_1, X_2, \ldots, X_n$ denote the portions of these geodesics between the given point and the spaces to which they are respectively perpendicular, then we have

$$X_r = \frac{R}{2} \log \frac{\sqrt{x^2 + x_r^2} + x_r}{\sqrt{x^2 + x_r^2} - x_r} \tag{11}$$

Consider the complete system of geodesics issuing from a particular point $(x_1^0, x_2^0, \ldots, x_n^0)$. It is represented by the following system of differential equations, the last of which is a consequence of the first,

$$\frac{dx_1}{x_1 - x_1^0} = \frac{dx_2}{x_2 - x_2^0} = \cdots = \frac{dx_n}{x_n - x_n^0} = \frac{dx}{x - \frac{z}{x}}$$

where we set

$$z = a^2 - x_1x_1^0 - x_2x_2^0 - \cdots - x_nx_n^0$$

for brevity. Condition (10) gives the following differential equation for the $(n-1)$-dimensional space orthogonal to all these geodesics:

$$\frac{dz}{z} = \frac{dx}{x},$$

which integrates to

$$\frac{a^2 - x_1x_1^0 - x_2x_2^0 - \cdots - x_nx_n^0}{\sqrt{a^2 - x_1^2 - x_2^2 - \cdots - x_n^2}} = C. \tag{12}$$

Comparing this equation with (8), one sees that the space it defines is again the locus of points equidistant from the point $(x_1^0, x_2^0, \ldots, x_n^0)$ and, letting ρ be this constant distance, we have

$$C = \sqrt{a^2 - (x_1^0)^2 - (x_1^0)^2 - \cdots - (x_n^0)^2} \cdot \cosh \frac{\rho}{R} = x^0 \cosh \frac{\rho}{R}.$$

Because of the way it was obtained, equation (12) remains valid when the point $(x_1^0, x_2^0, \ldots, x_n^0)$ goes to infinity, i.e. when x^0 tends to zero and ρ to infinity, and we see in this case that the product $x^0 \cosh \frac{\rho}{R}$ tends to a finite limit, which cannot differ from that of the product $\frac{1}{2}x^0 e^{\rho/R}$. Writing $\rho' - \rho$ in place of ρ and letting the point $(x_1^0, x_2^0, \ldots, x_n^0)$ tend to infinity while ρ remains constant we find, in the limit, the equation

$$\frac{a^2 - x_1x_1^0 - x_2x_2^0 - \cdots - x_nx_n^0}{\sqrt{a^2 - x_1^2 - x_2^2 - \cdots - x_n^2}} = ke^{-\frac{\rho}{R}} \tag{13}$$

where

$$(x_1^0)^2 + (x_2^0)^2 + \cdots + (x_n^0)^2 = a^2,$$

and this equation represents a system of $(n-1)$-dimensional spaces which can be defined as the *orthogonal trajectories of all the geodesics which converge to the same point* $(x_1^0, x_2^0, \ldots, x_n^0)$ *at infinity.* The different trajectories are distinguished by the values of the parameter ρ, which expresses the *constant* distance between any one of them and the trajectory determined by $\rho = 0$. The constant k is given when a point on the latter trajectory is given.

We now show that the nature of the space just considered is such that, by limiting any portion of it and transporting this portion to a different position, we can always obtain its *superposition* on a corresponding portion of the same space. To conceive how this can take place, imagine ∞^n points of space, infinitely close to each other, and connected in pairs by small geodesic arcs which measure their mutual distances.

Superimposability then consists in being able to disseminate the points in any other part of the space in such a way that they have the same disposition and mutual distances as the points considered initially. The n-fold net formed by the lines joining contiguous points in the initial portion can then be completely identified with the analogous net of the other portion without the connections between any points being broken or duplicated. The alterations undergone by the initial net when it is identified with the second may be no longer merely *apparent* when one considers the two nets to lie *in a space of more than n dimensions*: as long as this does not happen, however, the two nets have the character of *equality* by *congruence* or *symmetry.* The latter observation is related to an ingenious reflection of MOEBIUS.[14]

Suppose first that the space is referred to a new system of geodesic axes, of $y_1, y_2, \ldots, y_n$, with the same origin as the original and orthogonal to each other. Since all geodesics are represented by linear equations, it is clear that the transformation from the x variables to the y variables must be *linear*; and it is easy to convince oneself, in addition, that it must be of the form we have called *orthogonal.* In fact, the form (8) shows that the distance from the origin to any point $(x_1, x_2, \ldots, x_n)$ depends only on the function $x_1^2 + x_2^2 + \cdots + x_n^2$. We therefore have

$$x_1^2 + x_2^2 + \cdots + x_n^2 = y_1^2 + y_2^2 + \cdots + y_n^2$$
$$dx_1^2 + dx_2^2 + \cdots + dx_n^2 = dy_1^2 + dy_2^2 + \cdots + dy_n^2,$$

and hence

$$\frac{dx^2 + dx_1^2 + \cdots + dx_n^2}{x^2} = \frac{dy^2 + dy_1^2 + \cdots + dy_n^2}{y^2}.$$

This agreement in the form of the two elements shows clearly that two nets in which the corresponding vertices satisfy the equations

$$x_1 = y_1, \quad x_2 = y_2, \quad \ldots, \quad x_n = y_n$$

will be perfectly superimposable. But it is clear that the second of these nets is none other than the result of turning the first by a rotation about the origin which brings the original axes to the positions of the new ones. It is therefore proved that the superimposability of which we have spoken takes place by a simple rotation about the origin. Moreover, since we can more generally take

$$x_1 = \pm y_1, \quad x_2 = \pm y_2, \quad \ldots, \quad x_n = \pm y_n$$

with the freedom to combine signs in any way, it is clear that we have equality not only by *congruence* but also by several kinds of *symmetry.*

Since change of axes with fixed origin does not alter the form of the line element, it now remains to investigate the effect of a change of origin. And since, when choosing an arbitrary point in space, we can assume that the x_1 axis is directed through that point, it is permissible to take the new origin on that axis at the point $x_1 = a_1$. The new transformation therefore consists in preserving the x_1 axis and the preceding system of $x_2, \ldots, x_n$ coordinates, and replacing the system of geodesic perpendiculars to the space $x_1 = 0$

[14] *Der barycentrische Calcul*, p. 184 (Leipzig, 1827).

by the geodesic perpendiculars to the space $x_1 = a_1$, among which is the primitive x_1 axis. We call the new coordinates $y_1, y_2, \ldots, y_n$ and let b be a constant which plays the same rôle for them as a does for the x coordinates. We let $Y_1, Y_2, \ldots, Y_n$ be the geodesics analogous to $X_1, X_2, \ldots, X_n$ and we evidently have, as in (11)

$$Y_r = \frac{R}{2} \log \frac{\sqrt{y^2 + y_r^2} + y_r}{\sqrt{y^2 + y_r^2} - y_r}.$$

We then observe, leaving invariant the primitive systems of $x_2, x_3, \ldots, x_n$, that we have $X_r = Y_r$ for these systems, and consequently

$$\frac{x_r}{x} = \pm \frac{y_r}{y}, \quad r = 2, 3, \ldots, n. \tag{14}$$

By squaring and summing, first these equations, and then their differentials, we obtain the two formulae

$$\begin{gathered}(a^2 - x_1^2)y^2 = (b^2 - y_1^2)x^2 \\ \frac{\Omega^2}{x^2} + \left(d\frac{a}{x}\right)^2 - \left(d\frac{x_1}{x}\right)^2 = \frac{\Theta^2}{y^2} + \left(d\frac{b}{y}\right)^2 - \left(d\frac{y_1}{y}\right)^2\end{gathered} \tag{15}$$

where $\Theta^2 = dy^2 + dy_1^2 + \cdots + dy_n^2$. Secondly, if we consider the portions X_1^0, Y_1^0 on the x_1 axis between the two origins and the point where the axis is cut by the space $x_1 = x_1$, we have

$$X_1^0 = \frac{R}{2} \log \frac{a + x_1}{a - x_1}, \quad Y_1^0 = \frac{R}{2} \log \frac{b + y_1}{b - y_1}$$

while the distance between the two origins is

$$\frac{R}{2} \log \frac{a + a_1}{a - a_1}.$$

Then it is clear that we must put

$$X_1^0 = Y_1^0 + \frac{R}{2} \log \frac{a + a_1}{a - a_1},$$

i.e.

$$\frac{(a + x_1)(a - a_1)}{(a - x_1)(a + a_1)} = \frac{b + y_1}{b - y_1},$$

whence

$$y_1 = \frac{ab(x_1 - a_1)}{a^2 - a_1 x_1}, \quad x_1 = \frac{a(ay_1 + a_1 b)}{ab + a_1 y_1}. \tag{16}$$

These two formulae can be replaced by the relations

(17)
$$a^2 - x_1^2 = \frac{a^2(a^2 - a_1^2)(b^2 - y_1^2)}{(ab + a_1 y_1)^2}, \quad b^2 - y_1^2 = \frac{b^2(a^2 - a_1^2)(a^2 - x_1^2)}{(a^2 - a_1 x_1)^2},$$

which, suitably combined with the first of the equation (16), lead to the following two

$$\frac{a}{x}\sqrt{a^2 - a_1^2} = a\frac{b}{y} + a_1\frac{y_1}{y},$$
$$\frac{x_1}{x}\sqrt{a^2 - a_1^2} = a_1\frac{b}{y} + a\frac{y_1}{b},$$

whence

$$\left(d\frac{a}{x}\right)^2 - \left(d\frac{x_1}{x}\right)^2 = \left(d\frac{b}{y}\right)^2 - \left(d\frac{y_1}{y}\right)^2.$$

By virtue of the latter equation, the second of the equations (16) gives

$$\frac{dx^2 + dx_1^2 + \cdots + dx_n^2}{x^2} = \frac{dy^2 + dy_1^2 + \cdots + dy_n^2}{y^2},$$

hence the expression for the line element is again preserved under change of origin and consequently, by the above argument, superimposability holds in each case, since it now suffices to make an orthogonal transformation in order to render the new axes completely independent of the old.

Equations (14), (15, first), (17) give

$$x_r = \pm\frac{ay_r\sqrt{a^2 - a_1^2}}{ab + a_1 y_1}, \qquad r = 2, 3, \ldots, n,$$

from which we conclude, by (16, second) that the most general transformation of axes takes place by means of a *homography*.

Leaving aside this transformation of the coordinates $x_1, x_2, \ldots, x_n$ into others of the same kind, we can find another transformation which gives the element a remarkable form. This transformation, which may be called *polar*, is obtained by first putting

$$x_1 = r\lambda_1, \quad x_2 = r\lambda_2, \quad \ldots, \quad x_n = r\lambda_n$$

with the condition $\lambda_1^2 + \lambda_2^2 + \cdots + \lambda_n^2 = 1$. One obtains

$$dx_1^2 + dx_2^2 + \cdots + dx_n^2 = dr^2 + r^2 d\Lambda^2$$

where $d\Lambda^2 = d\lambda_1^2 + d\lambda_2^2 + \cdots + d\lambda_n^2$, hence

$$ds^2 = \left(\frac{Radr}{a^2 - r^2}\right)^2 + \frac{R^2 r^2}{a^2 - r^2} d\Lambda^2.$$

But if we let ρ be the geodesic distance from the origin, or pole, to the point $(x_1, x_2, \ldots, x_n)$, then we have

$$\frac{Radr}{a^2 - r^2} = d\rho, \quad \frac{r^2}{a^2 - r^2} = \sinh^2 \frac{\rho}{R}$$

hence

$$ds^2 = d\rho^2 + (R \sinh \frac{\rho}{R})^2 d\Lambda^2, \tag{18}$$

which is the justification for the name *polar*, because the variables are the radius ρ and the quantities λ which define the direction of the radius vector.

We can pass easily from this form to another which may be called *stereographic*, obtained by putting

$$\xi_r = 2R \tanh \frac{\rho}{2R} \cdot \lambda_r,$$

where ρ and λ_r have the above meaning. We derive

$$\lambda_r d\rho + R \sinh \frac{\rho}{R} \cdot d\lambda_r = d\xi_r \cdot \cosh^2 \frac{\rho}{2R},$$

$$\cosh^2 \frac{\rho}{2R} = \frac{1}{1 - \frac{\xi_1^2 + \xi_2^2 + \cdots + \xi_n^2}{4R^2}},$$

whence, squaring and summing the equations which result from the penultimate one by taking $r = 1, 2, \ldots, n$ and taking into account the last equation as well as (19), we have

$$ds = \frac{\sqrt{d\xi_1^2 + d\xi_2^2 + \cdots + d\xi_n^2}}{1 - \frac{\xi_1^2 + \xi_2^2 + \cdots + \xi_n^2}{4R^2}}. \tag{19}$$

This form was given without proof by RIEMANN, in the posthumous memoir cited above (II, §4).

RIEMANN has indicated another coordinate system from which he derives the curvature of a given space at a point (II, §2). These coordinates are, in certain respects, analogous to orthogonal cartesian coordinates, because they are derived from polar coordinates by putting

$$z_1 = \rho\lambda_1, \quad z_2 = \rho\lambda_2, \quad \ldots, \quad z_n = \rho\lambda_n.$$

We deduce

$$d\lambda_r = \frac{\rho dz_r - z_r d\rho}{\rho^2},$$

hence, squaring and summing,

$$d\Lambda^2 = \frac{(z_1^2 + z_2^2 + \cdots + z_n^2)(dz_1^2 + dz_2^2 + \cdots + dz_n^2) - (z_1 dz_1 + z_2 dz_2 + \cdots + z_n dz_n)^2}{\rho^4}$$

or

$$d\Lambda^2 = \frac{\sum(z_1 dz_2 - z_2 dz_1)^2}{\rho^4}$$

where the $\sum$ sign embraces all combinations of two indices. We also have

$$d\rho^2 = dz_1^2 + dz_2^2 + \cdots + dz_n^2 - \frac{\sum(z_1 dz_2 - z_2 dz_1)^2}{\rho^2},$$

whence, by substituting in (18), we finally obtain

(20)

$$ds^2 = dz_1^2 + dz_2^2 + \cdots + dz_n^2 + \frac{1}{\rho^2}\left[\left(\frac{R}{\rho}\sinh\frac{\rho}{R}\right)^2 - 1\right]\sum(z_1 dz_2 - z_2 dz_1)^2,$$

or

(20′)

$$ds^2 = dz_1^2 + dz_2^2 + \cdots + dz_n^2 + \frac{1}{3R^2}\left(1 + \frac{2\rho^2}{15R^2} + \cdots\right)\Sigma(z_1 dz_2 - z_2 dz_1)^2,$$

where $\rho^2 = z_1^2+z_2^2+\cdots+z_n^2$ and the convergent series in parentheses proceeds in increasing powers of $\frac{\rho}{R}$. For very small values of ρ we can simply take

$$ds^2 = dz_1^2 + dz_2^2 + \cdots + dz_n^2 + \frac{1}{3R^2}\Sigma(z_1 dz_2 - z_2 dz_1)^2.$$

Now, considering a surface element passing through the origin, we can arrange (by suitable choice of the axes $z_1, z_2, \ldots$ or $x_1, x_2, \ldots$) that this element coincides with that of the *surface* $z_3 = 0$, $z_4 = 0$, ..., $z_n = 0$, which has, in the neighbourhood of the origin, the line element

$$ds^2 = dz_1^2 + dz_2^2 + \frac{1}{3R^2}(z_1 dz_2 - z_2 dz_1)^2;$$

and, since the area of the infinitesimal triangle with vertices at the points $(0,0)$, (z_1, z_2), (dz_1, dz_2), the *second* of which is infinitely close to the origin, equals $\frac{1}{2}(z_1 dz_2 - z_2 dz_1)$, we conclude that $\sum(z_1 dx_2 - z_2 dz_1)^2$ equals four times the square of the area of the infinitesimal triangle with vertices at the points $(z_1, z_2, \ldots, z_n)$, $(dz_1, dz_2, \ldots, dz_n)$, the *second* of which is infinitely close to the origin. Thus if we divide the sum of fourth order terms in $(21)'$ by the square of the area of the infinitesimal triangle in question, we obtain the quotient $\frac{4}{3R^2}$; then, according to RIEMANN'S definition, this quotient, multiplied by $-\frac{3}{4}$, expresses the measure of curvature in the sense of the surface element considered, and shows that in this space the measure is constant and equal to $-\frac{1}{R^2}$ in all directions at any point.[15] For this reason it is proper to call this space a *space of constant curvature.*

A fourth transformation, the most important, is obtained by introducing n new independent variables $\eta, \eta_1, \ldots, \eta_{n-1}$ as follows:

$$\frac{Rx}{a - x_n} = \eta, \qquad \frac{Rx_1}{a - x_n} = \eta_1, \qquad \ldots, \qquad \frac{Rx_{n-1}}{a - x_n} = \eta_{n-1}.$$

One deduces immediately that

$$ds = \frac{\sqrt{d\eta^2 + d\eta_1^2 + \cdots + d\eta_{n-1}^2}}{\eta} \tag{21}$$

which implies incidentally that the formula (1) also represents the line element of a space of constant curvature when the $n+1$ variables $x, x_1, \ldots, x_n$ are independent and not subject to the relation (2), in which case the number of dimensions of the space is $n+1$ and the property of representability of geodesics by linear equations no longer holds.[16] A very remarkable consequence of the expression (21) is that $(n-1)$-dimensional space $\eta = \text{const.}$ has *zero* curvature at all points, because its line element has the form

$$ds = \text{const.} \sqrt{d\eta_1^2 + d\eta_2^2 + \cdots + d\eta_{n-1}^2}.$$

In fact, if we study the formula of RIEMANN, (19), we see immediately that the element cannot be reduced to the square root of a sum of squares of as many exact differentials as there are dimensions unless $\frac{1}{R} = 0$. The space

[15] To see the agreement between RIEMANN'S definition and that of GAUSS, we recall that, for GAUSS, the curvature of the surface defined by the element

$$ds^2 = d\rho^2 + m^2 d\theta^2$$

is $-\frac{1}{m}\frac{\partial^2 m}{\partial \rho^2}$, where m is in general a function of ρ and θ. If the variable ρ is the length of a geodesic arc issuing from a point of the surface at which the latter has an ordinary curvature, then the function is of the form $m = \rho(1 + m'\rho^2)$ where m' is a function which is neither zero nor infinity for $\rho = 0$ (see e.g. *Annali di Matematica*, ser. 2, I (1867), p. 358) and hence the measure of curvature at the point $\rho = 0$ equals $-6m'_0$. Then, the RIEMANN coordinates

$$z_1 = \rho\cos\theta, \quad z_2 = \rho\sin\theta$$

give the above element in the form

$$ds^2 = dz_1^2 + dz_2^2 + \frac{4(m^2 - \rho^2)}{\rho^4}\left(\frac{z_1 dz_2 - z_2 dz_1}{2}\right)^2$$

whence the RIEMANN measure of curvature at the point $\rho = 0$ is $-\frac{3}{4}\lim \frac{4(m^2-\rho^2)}{\rho^4}$. But $\lim_{\rho=0} \frac{m^2-\rho^2}{\rho^4} = 2m'_0$, so the two expressions coincide.

[16] The form (21) has been given, for the two-dimensional case only, by LIOUVILLE in his note on p. 600 of MONGE'S *Application de l'Analyse à la Géométrie* (Paris, 1850).

$\eta = 0$ is therefore one of those which RIEMANN calls *flat* (II, §1), among which are the plane and ordinary space, defined by the formulae

$$ds = \sqrt{dx^2 + dy^2}, \quad ds = \sqrt{dx^2 + dy^2 + dz^2}.$$

The equation $\eta = \text{const.}$ admits a very simple interpretation in the above terms. The point at infinity on the x_n axis has the coordinates

$$x_1 = x_2 = \cdots = x_{n-1} = 0, \quad x_n = a,$$

so that equation (13) becomes, for this point,

$$\frac{a - x_n}{x} = k' e^{-\frac{\rho}{R}}$$

where $k' = \frac{k}{a}$. Thus

$$\eta = \frac{R}{k'} e^{\frac{\rho}{R}}$$

and hence the equation $\eta = \text{const.}$ is equivalent to $\rho = \text{const.}$, whence we conclude (since the direction of the x_n axis is arbitrary) that the $(n-1)$-dimensional space $\eta = \text{const.}$ is none other than one of the orthogonal trajectories to the family of geodesics converging to a single point at infinity, i.e. to a system of geodesic *parallels*. Conversely, each of these orthogonal trajectories has zero curvature at all points, and hence any two of them (belonging to the same system, at least) will be superimposable on each other in all possible ways.

Introducing the variable ρ in place of η in (22), we obtain the equivalent form

$$(21') \qquad ds^2 = d\rho^2 + k'^2 e^{-\frac{2\rho}{R}} (d\eta_1^2 + d\eta_2^2 + \cdots + d\eta_{n-1}^2).$$

We have already seen that a set of $n-1$ linear equations in the coordinates $x_1, x_2, \ldots, x_n$ represents a geodesic. We shall now see, more generally, what is represented by a set of $n-m$ linear equations.

Assuming that these equations have been used to deduce expressions for $n-m$ of the coordinates as functions of the remaining m, it is evident that the number of independent parameters in such a system is $(m+1)(n-m)$. Imagine now that all the n coordinates $x_1, x_2, \ldots, x_n$ are expressed linearly as functions of the m variables $u_1, u_2, \ldots, u_m$. These expressions contain $(m+1)n$ parameters in all, but if we subject the parameters to an identity

$$x_1^2 + x_2^2 + \cdots + x_n^2 = u_1^2 + u_2^2 + \cdots + u_m^2 + h^2$$

(where h remains indeterminate) it is clear that we have adjoined $\frac{m(m+1)}{2} + m$ conditions,[17] so that the number of independent parameters remaining is

[17] Translator's note. Equating coefficients of the $u_i u_j$ and u_i after expressions for the x_k have been substituted in the left-hand side.

$(m+1)n - \frac{m(m+1)}{2} - m$. Now this number exceeds $(m+1)(n-m)$ by $\frac{m(m-1)}{2}$; hence the assumed relations between the x and u, together with the condition indicated, can always hold without restriction. We can then deduce from these relations, after putting

$$u^2 + u_1^2 + \cdots + u_m^2 = a^2 - h^2 = a'^2,$$

that

$$dx^2 + dx_1^2 + \cdots + dx_n^2 = du^2 + du_1^2 + \cdots + du_m^2$$
$$x^2 = u^2,$$

hence

$$ds = R\frac{\sqrt{du^2 + du_1^2 + \cdots + du_m^2}}{u},$$

with the condition

$$u^2 + u_1^2 + \cdots + u_m^2 = a'^2.$$

Consequently, the locus of points represented by the set of $n-m$ linear equations between the coordinates $x_1, x_2 \ldots, x_n$ is an n-dimensional space whose curvature is everywhere constant and equal to that of the original space.

For example, $n-2$ equations represent a *surface* of constant curvature (equal to $-\frac{1}{R^2}$), which it will be convenient to distinguish by the name *surface of first order*; $n-3$ equations represent a *space of three dimensions* with constant curvature (equal to $-\frac{1}{R^2}$); etc.

A real geodesic is determined without ambiguity by *two* points of the space; under the assumption in force so far there cannot be any exception to this property.

A surface of first order is determined without ambiguity by *three* points of the space. It entirely contains the geodesic through any two of its real points, hence, if two real surfaces of first order have two real points in common, they likewise have in common the geodesic determined by them.

A geodesic triangle is always situated on a unique surface of first order; the latter is also determined when the triangle is infinitesimal. This is because the geodesic prolongations of all the line elements in the same surface element have as locus a unique surface of first order.

When two surfaces of first order meet along a line, necessarily a geodesic, the angle between them is everywhere constant; i.e. if we take two line elements perpendicular to the intersection, one in the first surface and the other in the second, the infinitesimal distance between their extremities remains constant if their lengths remain constant. In fact,[18] if we assume

[18]The following proof, which can be omitted without loss of rigour, is inserted for the sake of the formulae to which it leads.

that the x_1 axis is directed along the common intersection of the two surfaces, the equations of the latter can evidently be put in the form

$$(x_2 = m_2 x_n, \quad x_3 = m_3 x_n, \quad \ldots, \quad x_{n-1} = m_{n-1} x_n),$$
$$(x_2 = m'_2 x_n, \quad x_3 = m'_3 x_n, \quad \ldots, \quad x_{n-1} = m'_{n-1} x_n),$$

where the m and m' are constant parameters. These two surfaces meet the space $x_1 = a_1$ along two geodesics which, by a preceding observation, are orthogonal to the x_1 axis. The two points with coordinates

$$(x_1 = a_1, \quad x_2 = m_2 x_n, \quad \ldots, \quad x_{n-1} = m_{n-1} x_n, \quad x_n = x_n),$$
$$(x_1 = a_1, \quad x_2 = m'_2 x'_n, \quad \ldots, \quad x_{n-1} = m'_{n-1} x'_n, \quad x_n = x'_n),$$

are situated on the first and second surface respectively, on precisely the two geodesics just mentioned, and their distance ρ is given, (8), by the formula

$$\cosh\frac{\rho}{R} = \frac{a^2 - a_1^2 - M x_n x'_n}{\sqrt{(a^2 - a_1^2 - m^2 x_n^2)(a^2 - a_1^2 - m'^2 x'^2_n)}},$$

where

$$m^2 = 1 + m_2^2 + \cdots + m_{n-1}^2, \quad m'^2 = 1 + m'^2_2 + \cdots + m'^2_{n-1},$$

$$M = 1 + m_2 m'_2 + \cdots + m_{n-1} m'_{n-1}.$$

Then, letting σ, σ' be the lengths of the two geodesics from the common point $x_1 = a_1$ to the two points considered, we find

$$\cosh\frac{\sigma}{R} = \frac{\sqrt{a^2 - a_1^2}}{\sqrt{a^2 - a_1^2 - m^2 x_n^2}}, \quad \cosh\frac{\sigma'}{R} = \frac{\sqrt{a^2 - a_1^2}}{\sqrt{a^2 - a_1^2 - m'^2 x'^2_n}},$$

whence

$$\sinh\frac{\sigma}{R} = \frac{m x_n}{\sqrt{a^2 - a_1^2 - m^2 x_n^2}}, \quad \sinh\frac{\sigma'}{R} = \frac{m' x'_n}{\sqrt{a^2 - a_1^2 - m'^2 x'^2_n}},$$

which shows that

$$\cosh\frac{\rho}{R} = \cosh\frac{\rho}{R}\cosh\frac{\sigma'}{R} - \frac{M}{mm'}\sinh\frac{\sigma}{R}\sinh\frac{\sigma'}{R}.$$

Since this formula no longer preserves any trace of the point a_1 taken on the x_1 axis, we see that, for any point of that axis in the two surfaces, the geodesics of lengths σ, σ' in the respective surfaces have a constant distance between their extremities. And, since this property holds for any lengths σ,

σ', it necessarily holds for infinitesimal lengths, whence comes the theorem we have enunciated.

Recalling, from what we have demonstrated above, that infinitesimal triangles are subject to all the relations of ordinary plane trigonometry, we recognise immediately, by making the lengths ρ, σ, σ' infinitesimal, that $\frac{M}{mm'}$ is the cosine of the angle between the initial line elements of the geodesics σ, σ', i.e. the angle between the two surfaces. On the other hand, it is easy to see that the triangle now considered can be an entirely arbitrary geodesic triangle. Thus the sides a, b, c and their opposite angles A, B, C in a geodesic triangle in the space considered satisfy the relation

$$\cosh\frac{a}{R} = \cosh\frac{b}{R}\cosh\frac{c}{R} - \sinh\frac{b}{R}\sinh\frac{c}{R}\cos A, \tag{22}$$

which differs from its analogue in spherical trigonometry only in the replacement of R by $R\sqrt{-1}$ (where R is the radius of the sphere). This fully agrees with the fact previously pointed out by MINDING,[19] and proved by CODAZZI[20] when one recalls that the geodesic triangle in question lies entirely on a surface of first order, i.e. of constant negative curvature, with respect to which it is also geodesic in the ordinary sense. If we assume that C is a right angle, then the two formulae which follow from (22) by permutation of the elements yield, when suitably combined,

$$\tanh\frac{a}{R} = \tanh\frac{c}{R}\cos B. \tag{23}$$

If we now imagine the vertex with angle A to recede indefinitely along the side b, while the side a remains fixed in position and length, then the hypotenuse c will grow indefinitely, and equations (22), (23) will give, in the limit

$$\cos A = 1, \quad \tanh\frac{a}{R} = \cos B.$$

The first formula shows that $A = 0$, i.e. that the two sides b, c approach asymptotically as the vertex with angle A goes to infinity. The second shows that the limit of the angle B is not a right angle, as in the plane, but an angle less than $90°$, whose magnitude depends on the distance a according to the formula

$$\tan\frac{B}{2} = e^{-\frac{a}{R}} \tag{24}$$

(equivalent to the preceding). If we call two geodesics converging to the same point at infinity *parallel*, as we have done previously, we see that we can draw *two* distinct geodesics parallel to a given geodesic through any

[19] *Journal für die reine und angewandte Mathematik* XX (1840), p. 323.

[20] *Annali di scienze matematiche e fisiche* (TORTOLINI) VIII (1857), p. 346.

point. These two parallels are equally inclined to the geodesic normal from the point to the given line, and their inclination B to the normal is related to the length a of that normal by the relation (24). This result fully agrees with the fundamental result of *noneuclidean geometry*, the principles of which, already familiar to GAUSS, have been summarised masterfully by LOBACHEVSKY[21] in synthetic form. The possibility of its construction by ordinary synthesis (limited to three-dimensional space) depends in the first place, as we have shown, on spaces of constant curvature (positive or negative) whose figures can be moved to arbitrary positions without undergoing any alteration in size or in the mutual disposition of contiguous elements, the *possibility* of which depends on the *existence of congruent figures* and hence on the *validity of the principle of superposition.* In the second place, in spaces of constant *negative* curvature, the geodesics are characterised, like euclidean lines, by the property of being uniquely determined by only *two* of their points, so they satisfy the *line axiom.* Likewise, the surfaces of first order are characterised, like euclidean planes, by the property of being determined by only three of their points, so they satisfy the *plane axiom.* Moreover, the relations between geodesics and surfaces of first order and those between surfaces themselves are the same as those between lines and planes and between planes themselves, since each such surface contains the geodesic between any two of its points, and two such surfaces meet along a geodesic (at a constant angle) if they meet at all. It follows from this correspondence that, if we admit the fundamental axioms of ordinary geometry, and exclude the parallel postulate, the theorems that we obtain are the same as those of the geometry of a space of constant negative curvature, since the latter geometry has the same basis as the former, with the exception of the postulate in question. The theorems of this geometry hold for all values of the curvature, which is a *parameter* of noneuclidean geometry (which I propose to call *pseudospherical* geometry), and it is *only by measurement* in the objective space that one can recognise whether its curvature is zero, i.e. $R = \infty$. Likewise it is *only by measurement* that one can assign a curvature to a given sphere; it is a *parameter* of spherical geometry.

In effect, one can verify that LOBACHEVSKY'S theory coincides, except in terminology, with the geometry of a three-dimensional space of constant negative curvature. Those who wish to see the development of this correspondence can find a more detailed exposition elsewhere.[22] Here, in order to avoid a long digression I confine myself to some brief indications.

Noneuclidean planimetry is nothing but the geometry of surfaces of constant negative curvature. The circles of this planimetry are lines which

[21] *Theory of Parallels.*

[22] See the *Essay on the interpretation of noneuclidean geometry*, whose particulars for the two-dimensional case can easily be repeated for three dimensions when the results of the present work are taken into account and recourse is made to an auxiliary space.

orthogonally cut all geodesic rays issuing from the same point of the surface, i.e. the geodesic circles. The circumference is given as a function of the geodesic radius r by the formula

$$\pi R\left(e^{\frac{r}{R}} - e^{-\frac{r}{R}}\right),$$

as was stated by GAUSS. Three points of the surface cannot always be connected by a geodesic circle with a real point as centre. The *horocycles* or *limit curves* of LOBACHEVSKY are nothing but geodesic circles with centre at infinity, whose radii thus form a system of geodesic parallels. Putting $n = 2$ in (21′), we have

$$ds^2 = d\rho^2 + k'^2 e^{-\frac{2\rho}{R}} d\eta^2$$

as the expression for the line element of a surface of constant negative curvature, referred to a system of concentric horocycles and their radii. The form of this expression shows that, with a suitable flexion of the surface, the horocycles can become the parallels of a surface of revolution whose meridian is the curve whose tangents are of constant length equal to R.[23]

Noneuclidean stereometry is nothing but geometry of three-dimensional spaces of constant negative curvature. We have already said what are the counterparts, in this geometry, of lines and planes. The counterparts of spherical surfaces are surfaces which orthogonally cut all geodesic rays issuing from the same point, i.e. the geodesic spheres. Three different points, and *a fortiori* four, cannot always be connected by a geodesic sphere with a real point as centre. The *horospheres* or *limit surfaces* of LOBACHEVSKY[24] are nothing but the geodesic spheres with centre at infinity, i.e. those whose radii form systems of geodesic parallels of the space of constant negative curvature. Putting $n = 3$ in (21) we have

$$ds = R\frac{\sqrt{d\eta^2 + d\eta_1^2 + d\eta_2^2}}{\eta} \tag{25}$$

where

$$\frac{Rx}{a - x_3} = \eta, \qquad \frac{Rx_1}{a - x_3} = \eta_1, \qquad \frac{Rx_2}{a - x_3} = \eta_2,$$

and conversely

$$x_1 = \frac{2aR\eta_1}{\eta^2 + \eta_1^2 + \eta_2^2 + R^2}, \qquad x_2 = \frac{2aR\eta_2}{\eta^2 + \eta_1^2 + \eta_2^2 + R^2},$$

$$x_3 = \frac{a(\eta^2 + \eta_1^2 + \eta_2^2 - R^2)}{\eta^2 + \eta_1^2 + \eta_2^2 + R^2}.$$

[23]Translator's note. That is, a pseudosphere.

[24]Or the F surfaces of Bolyai.

The formula (25) represents the line element of the noneuclidean space referred to a system of concentric horospheres and their radii. The form of this element shows that each horosphere, since it is represented by $\eta =$ const., is a surface of *zero* curvature, because its line element has the form

$$ds = \text{const.} \sqrt{d\eta_1^2 + d\eta_2^2},$$

and the variables η_1, η_2 are the *rectangular* coordinates of the points. A surface of first order

$$lx_1 + mx_2 + nx_3 + p = 0$$

is represented in the coordinates η, η_1, η_2 by the equation

$$2aR(l\eta_1 + m\eta_2) + (an + p)(\eta^2 + \eta_1^2 + \eta_2^2) = (an - p)R^2,$$

and hence it cuts the horosphere (for which $\eta =$ const.) along a circle. The latter reduces to a line only when $p = -an$, i.e. when the equation of the surface of first order has the form

$$lx_1 + mx_2 + n(x_3 - a) = 0,$$

in which case it is a *diametric* surface of the horosphere, passing through the centre (at infinity) of the latter. In this case the line of intersection is evidently a horocycle of the diametric surface, while on the horosphere it becomes a straight line when the latter is spread out on a plane. It follows that a triangle on the horosphere formed by three diametric surfaces is just a geodesic triangle on a surface of zero curvature, and hence it satisfies all the relations of ordinary plane trigonometry, since it likewise becomes a rectilinear triangle.

Thus all the concepts of noneuclidean geometry are perfectly matched in the geometry of a space of constant negative curvature. It remains to observe only that, whereas the concepts of planimetry receive a true and proper interpretation, because they are *constructible* on a *real* surface, those which embrace three dimensions are susceptible only to an analytic representation, because the space in which the representation could be realised is different from that to which we usually apply the name. At least, experience does not seem to accord with the results of this more general geometry unless we assume the constant R to be infinitely large, i.e. assume the space to be of zero curvature. It could be, however, that the triangles we have measured, and the portions of space we have observed, have been too small, just as measurements on a small portion of the terrestrial surface are insufficiently precise to reveal the sphericity of the globe.

Until now, we have spoken of n-dimensional spaces whose curvature is constant, but *negative*. The reason is that we have been mainly concerned with the reconciliation of these concepts with those of noneuclidean geometry, in comparison with which the opposite hypothesis is of minor interest. Nevertheless, we say a few words about it here.

The line element

$$ds = R\frac{\sqrt{dx_1^2 + dx_2^2 + \cdots + dx_n^2 - dx^2}}{x}, \tag{26}$$

where

$$x^2 = a^2 + x_1^2 + x_2^2 + \cdots + x_n^2,$$

belongs to an n-dimensional space whose curvature is constant and equal to $\frac{1}{R^2}$. This is obtained from (1) by changing R, a and x into $R\sqrt{-1}$, $a\sqrt{-1}$ and $x\sqrt{-1}$, and all the properties and equations derived analytically from the element (1) evidently persist, with the changes indicated, for the new element. For example, (8) becomes the following

$$\cos\frac{\rho}{R} = \frac{a^2 + x_1x_1^0 + x_2x_2^0 + \cdots + x_nx_n^0}{\sqrt{(a^2 + x_1^2 + \cdots + x_n^2)(a^2 + (x_1^0)^2 + \cdots + (x_n^0)^2)}}, \tag{27}$$

a formula which gives a real value for ρ for any real values of x_1, x_2, ..., x_n, x_1^0, x_2^0, ..., x_n^0. It is clear that the theorem about superimposability of two portions of space continues to hold for these spaces.

If we assume that the variables $x, x_1, \ldots, x_n$, and the constants R, a, in (26) are real, then there is no limit on the admissible values of the coordinates $x_1, x_2, \ldots, x_n$. They can vary from $-\infty$ to $+\infty$. For all real values of these coordinates the space is *continuous* and *simply connected, but not infinite* (RIEMANN III, §2), because if we substitute

$$x_1^0 = \lambda_1\tau, \quad x_2^0 = \lambda_2\tau, \quad \ldots, \quad x_n^0 = \lambda_n\tau,$$

where $\lambda_1^2 + \lambda_2^2 + \cdots + \lambda_n^2 = 1$ in (27) we have, for $\tau = \infty$,

$$\cos\frac{\rho}{R} = \frac{\lambda_1x_1 + \lambda_2x_2 + \cdots + \lambda_nx_n}{\sqrt{a^2 + x_1^2 + x_2^2 + \cdots + x_n^2}},$$

a formula which gives a determinate and finite value for ρ. Geodesics continue to be represented by linear equations; however, on account of the admissibility of infinite values of the coordinates, the principle that *two* points determine a geodesic without ambiguity *ceases* to be true *without restriction*. In fact, let

$$x_1 = b_1x_n + b_1', \quad x_2 = b_2x_n + b_2', \quad \ldots$$

be the equation of a geodesic. As long as at least one of the points known on it has finite coordinates, the coefficients can be determined without ambiguity. But if both the points have infinite coordinates we have to put the equations into the form

$$\frac{x_1}{x_n} = b_1 + \frac{b_1'}{x_n}, \quad \frac{x_2}{x_n} = b_2 + \frac{b_2'}{x_n}, \quad \ldots$$

and replace the first terms by the limits to which they converge at the two points. If these limits are equal in pairs, the values of the second coordinates remain indeterminate, and the geodesic is *not* unique. If the limits are different, all points of the geodesic have infinite coordinates.

The considerations that led us to equation (13) are not applicable to spaces of constant positive curvature, because the latter have no points at infinity. The figures represented by this equation therefore have no counterparts in the new spaces, just as there are no geodesic *parallels*.

One sees that the geometry of spaces of constant positive curvature (which can appropriately be called *spherical geometry* in the broad sense, since, as equation (22) shows, the geodesic triangles are subject to the laws of spherical trigonometry) differs very markedly from *pseudospherical* geometry, even though both admit congruent figures. Moreover, pseudospherical geometry leads spontaneously to the consideration of spaces of positive curvature. In fact, if we put

$$\frac{a}{x} = y, \qquad \frac{x_1}{x} = y_1, \qquad \ldots, \qquad \frac{x_n}{x} = y_n$$

in (26), we find

$$ds = R\sqrt{dy^2 + dy_1^2 + \cdots + dy_n^2}$$

with the condition

$$y^2 + y_1^2 + \cdots + y_n^2 = 1,$$

which, when we compare with equation (18) and take $\rho = \text{const.}$, tells us that the geodesic spheres of radius ρ, in an n-dimensional space of constant negative curvature $-\frac{1}{R^2}$, are the $(n-1)$-dimensional spaces of constant curvature $\left(\frac{1}{R\sinh\frac{\rho}{R}}\right)^2$. Thus spherical geometry can be regarded as a part of pseudospherical geometry.

Bologna, August 1868.

Translator's Introduction

Klein's *On the so-called noneuclidean geometry*

Felix Klein began his mathematical career as assistant to the physicist and geometer Julius Plücker, between 1866 and 1868. Under Plücker's influence, Klein developed a strong belief that the fundamental concepts of geometry are projective, and that projective geometry is best pursued by algebraic methods, in particular, homogeneous coordinates. It is clear in all his writings, from this paper to his posthumously published book *Vorlesungen über Nicht-euklidische Geometrie* (Klein [1928]), that Klein believes in points at infinity, so for him "the plane" is *not* the euclidean plane but the projective plane.

This is a perfectly natural belief, since our visual world is projective rather than euclidean. Objects constantly change shape as we shift our point of view – circles become ellipses, right angles become acute or obtuse, and so on. We are also used to seeing points at infinity; they are the points on the horizon, and they form a line. Yet it is also clear that, somehow, we abstract euclidean geometry from the visual world of projective images. For example, we immediately interpret Figure 3.1 as a tiling of the plane by identical rectangles, even though the angles are not right angles and the tiles decrease in size.

Thus experience suggests that euclidean geometry exists within projective geometry, but it is not entirely trivial to explain this intuition. As Klein mentions in the preamble to his paper, it was first done by Cayley [1859]. Cayley also gave a construction of spherical geometry, but he overlooked the possibility of hyperbolic geometry, which was Klein's decisive contribution. Klein gave a general method of constructing length and angle measures in projective geometry, and showed that three essentially different geometries result: spherical (or *elliptic*), euclidean (or *parabolic*), and *hyperbolic.*

Klein completed his investigation by defining *curvature* of measures, showing that it coincides with Gaussian curvature and that the three geometries correspond to the respective cases of constant positive, zero, and

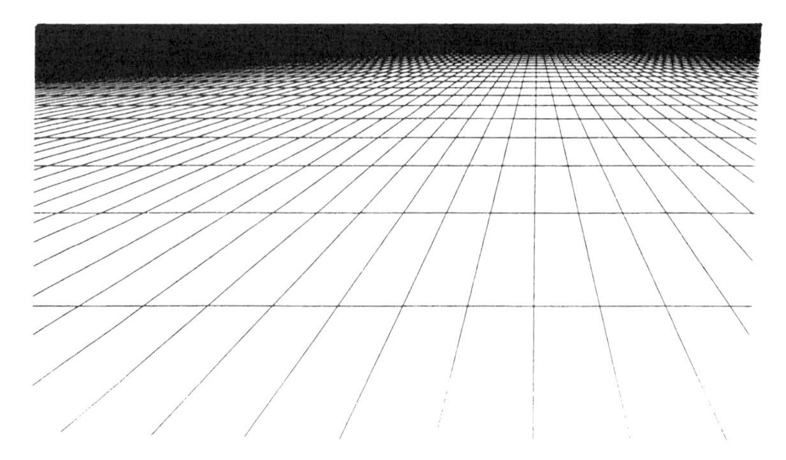

Figure 3.1: Tiling of the plane

negative curvature. This is similar to the path followed by Beltrami, except that Beltrami proceeded in the opposite direction and did not cover the full distance from Gaussian curvature to the projective plane. He stopped at the euclidean plane. In particular, his model of the hyperbolic plane lies in a finite part of the euclidean plane – the unit disc – which is quite sufficient for his purposes, since points outside the disc are not part of the model. (They do throw more light on the model, however. As Klein points out, they enable each motion of the hyperbolic plane to be regarded as a rotation: there are "elliptic" rotations about points inside the disc, "parabolic" rotations about boundary points of the disc, and "hyperbolic"rotations about points outside the disc.)

The difference between the Klein and Beltrami models may seem small, but it was significant in an era when the foundations of geometry were not located where they are today. Beltrami's construction showed that hyperbolic geometry is consistent if euclidean geometry is consistent. Klein's, on the other hand, showed that spherical, euclidean and hyperbolic geometry are *all* consistent if projective geometry is consistent. Thus Klein had a more comprehensive result, and technically a superior one, since projective geometry is logically prior to euclidean geometry (thanks to Cayley's construction). Beltrami was satisfied with euclidean geometry as a foundation, but Klein built on the deeper foundation of projective geometry.

In fact, the consistency of hyperbolic geometry relative to euclidean was only one of a series of relative consistency results obtained by Klein. In §15 he constructs elliptic and hyperbolic geometry from euclidean geometry, and euclidean geometry from elliptic. The independence of the parallel axiom was merely an afterthought. He mentioned it only in his second paper on noneuclidean geometry (Klein [1873]). I have appended a translation of the relevant passage to the paper below.

Today, the superiority of Klein's foundation seems only marginal. Since the arithmetisation of analysis in the late 19th century and Hilbert's investigation of the foundations of geometry (Hilbert [1899]), the theory of real numbers has been adopted as the most convenient foundation for all geometries. "Points" are defined as n-tuples of real numbers, and "lines," "planes" etc. are defined by appropriate equations. Different geometries are distinguished by different definitions of "distance" or measure. Klein himself made a large contribution to foundational studies with his Erlanger Programm (Klein [1872]), according to which a geometry is defined by a space and a *group of transformations.* The germ of his idea can be seen in the present paper, where measures are defined as quantities invariant under particular groups of linear transformations.

It follows from Hilbert's construction of geometry that spherical, euclidean and hyperbolic geometry are all consistent provided the theory of real numbers is consistent. This removes the question of consistency from geometry altogether, though admittedly it does not answer it. In fact, Hilbert's strategy of reducing consistency problems in mathematics to problems about numbers was eventually thwarted by Gödel's theorem (Gödel [1931]) that the consistency of number theory is *not* provable within number theory itself. This profound and astonishing result shows that in some sense it is appropriate to stop trying to prove hyperbolic geometry consistent. We know it is as consistent as euclidean geometry, and *at least* as consistent as the theory of real numbers, but further than that we cannot go.

Klein's exposition of his approach to geometry is so clear that little introduction is necessary. He cleverly motivates the key concepts by beginning with the 1-dimensional case (§3), which is simple but not too simple. Considering only the transformation $z \mapsto \lambda z$, he shows how to construct a "scale" of points on the line "equally spaced" with respect to this transformation, and how this leads naturally to the use of logarithms and cross-ratios in defining "distance." Over the next few sections he develops a thorough understanding of measures on the line (and the similar concept of angle measure), after which it is relatively easy to construct 2-dimensional (§8) and 3-dimensional (§16) measures and the corresponding geometries.

A curious feature of the construction is that the euclidean measure actually requires more work than the other two. The general formula for measure contains an arbitrary constant c which is not important for spherical or hyperbolic geometry, but crucial in the euclidean case. For a fixed finite value of c, the euclidean formula makes all distances zero, and it is only by allowing c to tend to infinity in a certain way that a nontrivial measure is obtained. As expected, it turns out to be the standard euclidean measure, which gives distance $\sqrt{(x_2 - x_1)^2 + (y_2 - y_1)^2}$ between points (x_1, y_1) and (x_2, y_2). In retrospect, this peculiarity of the euclidean case is understandable because it *is* a limiting case (as curvature tends to zero).

The main disappointment in Klein's paper, for those who have seen his later work, is the complete absence of pictures. Readers are encouraged to look at Fricke and Klein [1897] and Klein [1928] to find a wealth of illustrations. Here are just a few, suitable for the present paper. The first two illustrate Klein's basic construction of noneuclidean scales on a line.

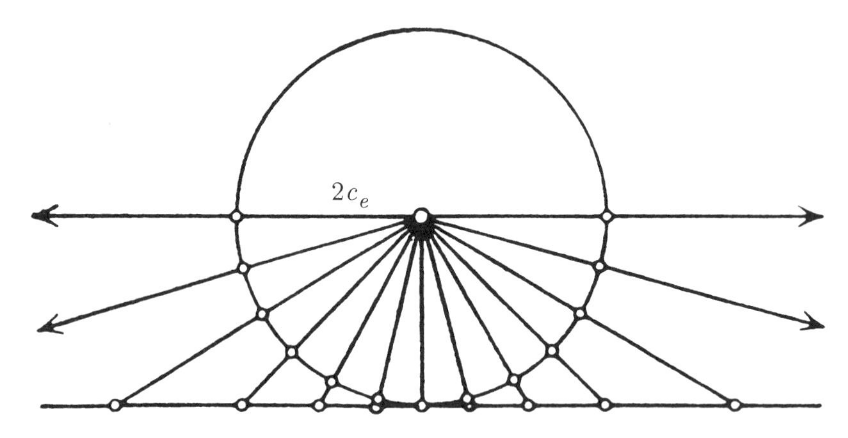

Figure 3.2: Construction of a scale of elliptically equidistant points on the line (Klein [1928], Figure 109; reprinted with permission from Springer-Verlag).

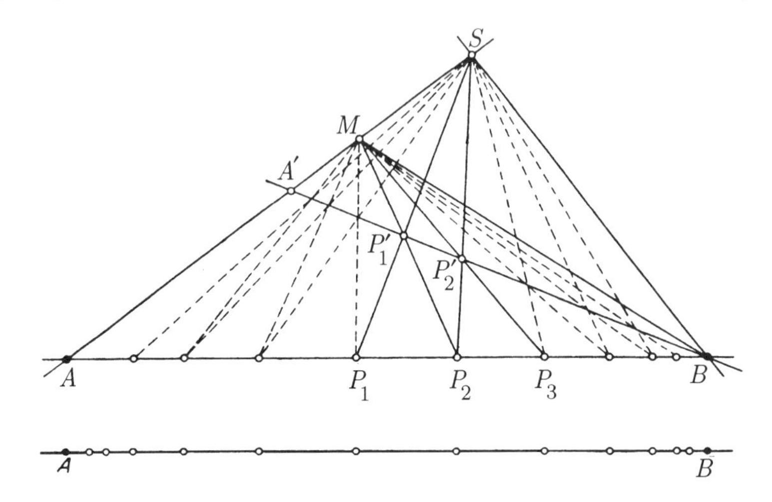

Figure 3.3: Hyperbolically equidistant points on the line (Klein [1928], Figure 112; reprinted with permission from Springer-Verlag).

The next two are pictures of the hyperbolic plane, modelled by the interior of an ellipse. The first contains a grid formed by a family of horocycles and the family of lines through its center at infinity. The second shows the effect of a translation, with a line and its equidistant curves mapped into themselves.

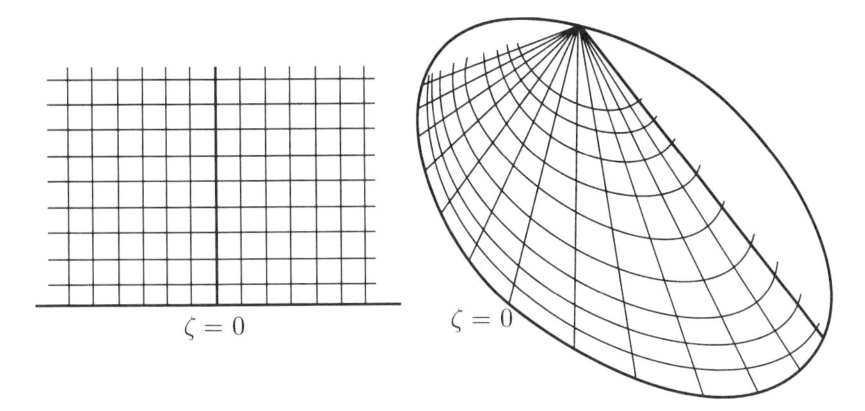

Figure 3.4: Grid of euclideanly equidistant points within the ellipse (Fricke and Klein [1897], Figure 1).

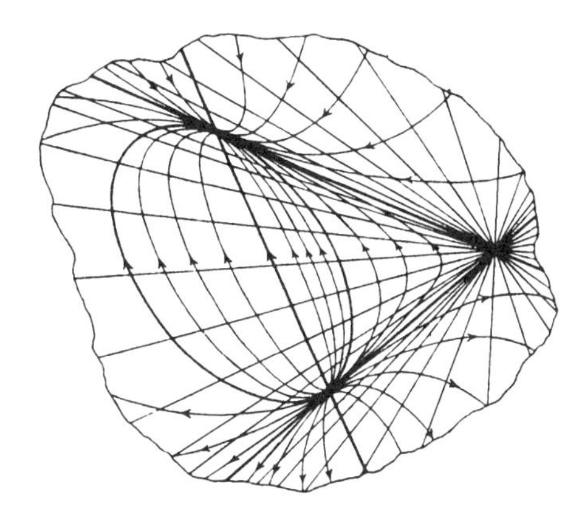

Figure 3.5: A hyperbolic motion of the ellipse, viewed as rotation about a point outside the ellipse (Fricke and Klein [1897], Figure 3).

References

A. Cayley [1859], A sixth memoir on quantics, *Phil. Trans. Roy. Soc.* **149**, 61–90.

R. Fricke and F. Klein [1897], *Vorlesungen über die Theorie der Automorphen Functionen*, B. G. Teubner, Leipzig.

K. Gödel [1931], Über formal unentscheidbare Sätze der Principia Mathematica und verwandte Systeme, *Monat. Math. Phys.* **38**, 173–198.

D. Hilbert [1899], *Grundlagen der Geometrie*, B. G. Teubner, Leipzig, 1899. English translation: *Foundations of Geometry*, Open court, Chicago, 1971.

F. Klein [1872], *Vergleichende Betrachtungen über neuere geometrische Forschungen (Erlanger Programm)*, Akademische Verlagsgesellschaft, Leipzig, 1974.

F. Klein [1873], Über die sogenannte Nicht-Euklidische Geometrie, *Math. Ann.*, **6**, 112–145.

F. Klein [1928], *Vorlesungen über Nicht-Euklidische Geometrie*, Springer, Berlin.

On the so-called noneuclidean geometry[25]

by Felix Klein

Mathematische Annalen 4 (1871) 573–625

The following discussion concerns the so-called noneuclidean geometry of Gauss, Lobachevsky and Bolyai, and the related reflections of Riemann and Helmholtz on the foundations of our geometric concepts. We shall not further pursue the philosophical speculations which led to the works in question; rather, it is our purpose to *present the mathematical results of these works, insofar as they relate to the theory of parallels, in a new and intuitive way, and to provide a clear general understanding.*

The route to this goal is through projective geometry. By the results of Cayley,[26] one may construct a projective measure on ordinary space using an arbitrary second degree surface as the so-called fundamental surface. Depending on the type of second degree surface used, this measure will be a model for the various theories of parallels in the above-mentioned works. But it is not just a model for them; as we shall see, it precisely captures their inner nature.

I shall begin with a brief description of the various theories in question (§1). Then I shall turn to the Cayley measure, which will be developed with the various theories of parallels constantly in view. This calls for a detailed discussion, as these investigations of Cayley do not seem to be well known, and also because his viewpoint is different from ours. Cayley is concerned with showing that ordinary (euclidean) geometry may be regarded as a special part of projective geometry. For this purpose he sets up the

[25] Cf. a note with the same title in the *Göttinger Nachrichten*, 1871.

[26] In the Sixth Memoir upon Quantics, *Phil. Transactions* 149, 1859 and *Coll. Papers* II, pp. 583-592. Cf. Fiedler's translation of Salmon's *Conic Sections*, 2nd Ed. (Leipzig 1866), or Fiedler: *Die Elemente der neueren Geometrie und der Algebra der binären Formen* (Leipzig 1862).

general projective measure and then shows that its formulae become the formulae of ordinary geometry when the fundamental surface degenerates to a particular conic section, the imaginary circle at infinity. Our goal, on the other hand, is to explain the *geometric content* of the general Cayley measure as clearly as possible, and to recognise, not only how euclidean geometry results from its specialisation, but more importantly that it is related in exactly the same way to the various metric geometries arising from the different theories of parallels.

This discussion yields some new viewpoints. Among them I count the realisation of the Cayley measure by repeated space transformations. I would also emphasise the form under which the concept of curvature appears in §7 and §14.

Incidentally, the definition I give for the projective measure is somewhat more general than the one given by Cayley himself. In order to determine the distance between two points, I imagine them to be connected by a straight line. The latter cuts the fundamental surface in two other points, which have a certain cross-ratio with the given two. *The logarithm of this cross-ratio, multiplied by an arbitrary, but fixed, constant c, is what I call the distance between the two points.* In order to determine the angle between two planes, I take the two tangent planes to the fundamental surface through their line of intersection. The latter form a certain cross-ratio with the two given planes. *The logarithm of this cross-ratio, multiplied by a fixed constant c', is what I call the angle between the two given planes.* These geometric definitions agree with the analytic definitions given by Cayley as soon as one takes particular values for c and c', namely, both[27] equal to $\frac{\sqrt{-1}}{2}$. However, it is essential in what follows to keep c and c' open, since e.g. c corresponds to the characteristic constant of noneuclidean geometry.

§1
The different theories of parallels

As is well known, the eleventh axiom of Euclid is equivalent to the theorem that the angle sum of a triangle is two right angles. Legendre was able to show[28] that the angle sum of a triangle cannot be greater than two right angles. He also showed that when the angle sum equals two right angles in one triangle, then the same is true of all triangles. However, he was not able to show that the angle sum cannot be less than two right angles.

[27]Occasionally Cayley also takes the "quadrant" as unit. This amounts to taking both c and c' equal to $\frac{\sqrt{-1}}{\pi}$.

[28]This proof, like the one of Lobachevsky for the same result, assumes that lines are infinitely long. If this assumption is dropped (cf. the text below) then the proof fails, as one clearly sees that otherwise it would apply to geometry on the sphere.

Similar considerations seem to have been the starting point of Gauss's investigations in this subject. Gauss was of the opinion that it was impossible to prove the theorem that the angle sum is two right angles, and that a geometry could be constructed, satisfying the other axioms, but with smaller angle sums. He called this geometry *noneuclidean.*[29] He spent a lot of time on it but unfortunately published nothing, apart from a few hints. This noneuclidean geometry involves a certain characteristic constant for the spatial measure. If one gives it the value infinity, then one obtains ordinary euclidean geometry. But if the constant has a finite value, then one has a different geometry, for which the following laws are valid (among others):

The angle sum of a triangle is less than two right angles, and indeed by an amount proportional to the area of the triangle. For a triangle with vertices at infinity the angle sum is zero. Through a point outside a line one can draw two parallels to the line, i.e. lines which meet the given line at infinity on one side or the other. The lines through the point, but between the parallels, do not meet the given line at all.

This noneuclidean geometry was encountered by Lobachevsky,[30] Professor of Mathematics at the University of Kazan and, a few years later, by the Hungarian mathematician J. Bolyai,[31] and they made it the subject of detailed publications. However, these works remained largely unknown until the publication of the correspondence between Gauss and Schumacher, from 1862 onwards, drew attention to the subject. Since then the opinion has grown that the theory of parallels should now be recognised as genuinely indeterminate.

But this opinion has undergone an essential modification since the appearance of Riemann's inaugural lecture *Über die Hypothesen, welche der Geometrie zugrunde liegen* [On the hypotheses that lie at the foundations of geometry] in 1867, the year after his death, and the investigations of Helmholtz that soon followed in the Göttinger Nachrichten (1868, no. 6), *Über die Tatsachen welche der Geometrie zugrunde liegen* [On the facts that lie at the foundations of geometry].

Riemann's essay points out that the unboundedness of a space does not necessarily imply its infinitude. It is conceivable, and does not contradict our intuition, which only relates to a finite portion of space, that a space could be finite and curved around on itself. The geometry of our space would then

[29]Cf. Sartorius v. Waltershausen, *Gauss zum Gedächtnis*, p. 81. There are some letters on the subject in the correspondence between Gauss and Schumacher.

[30]In the Kazan reports of 1829, 1835-8, *Crelle's Journal*, 17, 1837 (Géometrie imaginaire), Geometrische Untersuchungen zur Theorie der Parallellinien, Berlin 1840, Pangéométrie, Kazan 1855. (An Italian translation of the *Pangéométrie* may be found in vol. 5 of the *Giornale di Matematiche*, 1867.) [Coll. Geom. Works, vols. I, II, Kazan 1883, 1886]

[31]In an appendix to W. Bolyai's *Tentamen juventutem* ... Maros-Vasarkely, 1832. Cf. the Italian translation in vol. 6 of the *Giornale di Matematiche*, 1868.

be like that of a 3-dimensional sphere in a 4-dimensional manifold. This idea, which also occurs in Helmholtz, carries with it triangles with angle sum greater[32] than two right angles (as in ordinary spherical triangles), and indeed exceeding it proportionally to the area of the triangle. Straight lines have no points at infinity, and indeed one cannot draw any parallel at all to a given line through a point outside it.

A geometry based on these ideas could be placed alongside ordinary euclidean geometry like the above-mentioned geometry of Gauss, Lobachevsky and Bolyai. While the latter gives each line two points at infinity, the former gives none at all (i.e. it gives two imaginary points). Between the two, euclidean geometry stands as a transitional case; it gives each line two coincident points at infinity.

Following the usual terminology[33] in modern geometry, these three geometries will be called *hyperbolic, elliptic* and *parabolic* in what follows, according as the points at infinity of a line are real, imaginary or coincident.

These three geometries will turn out to be special cases of the general Cayley measure. One obtains the parabolic (ordinary) geometry by letting the fundamental surface for the Cayley measure degenerate to an imaginary conic section. If one takes the fundamental surface to be a proper, but imaginary, surface of second degree, one obtains the elliptic geometry. Finally, the hyperbolic geometry is obtained when one takes the fundamental surface to be a real, but not ruled, surface of second degree and considers the points inside it.

I now turn to setting up the general Cayley measure, first for a base figure of the first kind. In the process I discuss how projective concepts subsume the concepts of elliptic and hyperbolic geometry.

It is worth mentioning, at this point, the connection between the geometric objects under consideration and the concept of a measure in arbitrary analytic manifolds.

It was Beltrami who first showed[34] how the planimetric part of hyperbolic (noneuclidean) geometry could be interpreted using the ordinary measure on surfaces with constant negative curvature. Thus in hyperbolic geometry the plane can be regarded as a 2-dimensional manifold of constant negative curvature. With the appearance of Riemann's work, for the first time extending the concept of curvature to higher dimensional manifolds,

---

[32] As already mentioned, the proofs of Legendre and Lobachevsky to the contrary depend on the assumption that space is infinite.

[33] E.g. one calls the points of a surface hyperbolic, elliptic or parabolic according as the principal tangents are real, imaginary or coincident. Steiner called an involution hyperbolic, elliptic or parabolic according as the double elements were real, imaginary or coincident, etc.

[34] Saggio di interpretazione della Geometria non-euclidea. *Giornale di Matematiche*, 1868, Beltrami's works, vol. I, 374-405.

Beltrami extended his investigations to spaces with an arbitrary number of dimensions.[35] In particular, he showed that the hyperbolic geometry of ordinary (3-dimensional) space may also be attributed to constant negative curvature, in fact the assumption of constant negative curvature is equivalent to the assumption of hyperbolic geometry. On the other hand, elliptical geometry – or spherical geometry as he calls it[36] (because ordinary spherical geometry belongs here) – results from constant positive curvature of space. Finally, parabolic geometry corresponds to curvature which is also constant, but zero.

Now since it will be shown that the general Cayley measure in space of three dimensions covers precisely the hyperbolic, elliptic and parabolic geometries, and thus coincides with the assumption of constant curvature, one is led to the conjecture that the general Cayley measure agrees with the assumption of constant curvature in any number of dimensions. This is in fact the case, though we shall not show it here. It allows one to use formulae, in any spaces of constant curvature, which are presented here assuming two or three dimensions. It includes the facts that, in such spaces, geodesics can be represented by linear equations, like straight lines,[37] and that the elements at infinity form a surface of second degree, etc. These results have already been proved by Beltrami, proceeding from other considerations;[38] in fact, it is only a short step from the formulae of Beltrami to those of Cayley.

At the same time, we shall indicate the connection between the following and the general investigations of Christoffel[39] and Lipschitz[40] on differential expressions.

§2
Generalities on space metrics

As is well known, all space measures go back to two fundamental problems: determination of the *distance between two points*, and determination of the *relative slope of two intersecting lines*. In fact the instruments of the practical geometer measure either *line segments* or *angles*; all other quantities can be computed from these.

[35]Teoria fundamentale degli spazii di curvatura costante. *Annali di Matematica*, ser. II, vol. 2, 1868/9. Beltrami's works, vol. I, 406-429.

[36]On the other hand, he calls hyperbolic geometry "pseudospherical".

[37]In particular, spaces of constant curvature admit projective geometry. Cf. §17 of the text.

[38]First for surfaces of constant curvature in an essay: Risolutione del problema di riportare i punti di una superfice etc., *Annali di Matematica*, ser. I, vol. 7, 1866. Works, vol. I, 262-280. Then for the general case in the above-mentioned work: Teoria generale etc.

[39]Borchardt's Journal, 70 (1869), p. 46.

[40]Borchardt's Journal, 70 (1869), p. 71, 72 (1870), p.1.

In the setting of projective geometry one can call both these basic problems the *problem of determining a measure on the basic figure of the first kind.* Measuring the distance between two points corresponds to determining the measure on the linear point series; measuring the relative slope of two intersecting lines corresponds to determining the measure on the planar pencil of lines. Finally, determining the measure on the pencil of planes is the same as for the planar pencil of lines, since the relative slope of two planes is taken to be the relative slope of two lines, in which the given planes meet a plane perpendicular to their line of intersection. Thus it suffices to determine the measure on the linear point series and on the planar pencil of lines, and this is where we shall begin.

As long as one considers the point series and the line pencil to lie in the plane they are related by the principle of duality. This is not the case for the corresponding measures, which are essentially different, e.g.:

The distance between two points is an algebraic function of their coordinates, whereas the angle between two lines is a transcendental (circular) function.

The length of an unbounded point series is infinite, whereas the total angle of a line pencil is finite.

A line segment is uniquely determined (up to sign), but an angle is determined only up to multiples of a period. The segment can be subdivided, in a simple fashion, into an arbitrary number of equal parts. Not so the angle, which in general admits bisection only, etc.

In spite of these differences the two kinds of measure have something in common, inasmuch as both may be subsumed under a more general measure. The common features are of two kinds.

First, both measures satisfy an addition law,[41] i.e. the measure of $\overline{12}$, plus the measure of $\overline{23}$, equals the measure of $\overline{13}$ – in symbols, $\overline{12}+\overline{23}=\overline{13}$. This *additivity of measure* is a general law, given from the beginning for all metrics on 1-dimensional manifolds.[42] In determining those functions of coordinates that can represent metrics, additivity serves as a functional equation. Along with additivity there is another property of measure which applies to all 1-dimensional manifolds, namely, that the distance of an element from itself is zero: $\overline{11}=0$. It follows from this and the property just mentioned that, in particular, $\overline{12}=-\overline{21}$.

Second, the measures considered here have another property, that makes them suitable for measurement in space. This property is that *they are not altered by a motion in space.* In particular, the angle between two lines in a pencil does not alter when one allows the pencil to rotate about its centre. Likewise, the distance between two points of a line does not alter when the

[41] Of course, with angle measure this only holds up to multiples of π.

[42] The same holds, e.g., when we measure time or weight or intensities.

line is displaced along itself.

The two properties mentioned are sufficient to characterise both measures; they also appear clearly in the means used for actual measurement. In both angle and length measurement, one uses a *scale of equidistant elements*, and lays it arbitrarily against the object to be measured.[43] The number of scale divisions between the two elements whose difference of measures is to be determined is the desired measure. We shall not discuss the fact that this number of scale divisions is not in general an integer or even rational, nor the fact that actual measurement is not exact but only within certain limits. However, we must examine more closely how the two properties mentioned are involved in the process of measurement. The first property, additivity, is immediately evident when we take the distance between two elements as the number of scale divisions between them. The second property is involved when we find the same number, no matter how the scale is laid against the object being measured. For this to happen the scale must have the property of fitting itself under an arbitrary motion within itself. In other words, if the scale is subjected to a motion, which leaves the underlying point series or line pencil unaltered and moves one scale division onto the next, then every scale division moves onto the next.

The latter property of the scale allows it *to be constructed by repeated motions* (as one in practice actually does).

In particular, to construct a scale on the linear point series one takes two points (1) and (2) as ends of the first scale division. Then one displaces the line along itself until (1) coincides with (2). This carries (2) to another point (3), which is the third scale division point. If one makes the same displacement again, then (1) is carried to (2), (2) to (3), and finally (3) to a new scale point (4), etc.

Similarly, if one wants a scale on the planar pencil of lines, one begins with two rays (1) and (2) as ends of the first scale division.[44] A rotation of the pencil about its centre carrying (1) to the position of (2) will carry (2) to a position (3) which is the third ray on the scale, etc.

Displacement of a point series or rotation of a line pencil in projective geometry both fall under the general concept of a *linear transformation which maps the basic figure onto itself.* This immediately gives a *more general construction of a scale* for the linear point series or the pencil of lines, and thereby a *more general measure* on these basic figures, which includes the actual constructions of measures as special cases. Namely, one repeatedly

[43]For line measurement one uses a scale of equidistant points on a line, as asserted in the text, a *ruler.* However, to measure angles one does not use an angle scale, but a *divided circle* in its place. The idea of an angle scale will be retained in the text, however, because the circle is not a basic figure in the sense of projective geometry.

[44]In practice, one takes the scale division to be an angle such that the right angle is expressible as an integral number of scale divisions, but this is not a consideration here.

applies a linear transformation to an element of the figure for which the scale is required, in such a way as to map the given figure into itself. The initially chosen element then generates a series of elements which make up the scale. The distance between two elements is then the number of scale divisions between them.[45] Thus in the first instance distance is defined only between elements separated by an integral number of scale divisions, but by continued subdivision of the scale (cf. the paragraphs below) it is also possible to define the distance between two elements separated by a rational number of scale divisions. Finally, by admitting the concept of irrational limit, one is able to speak of the distance between arbitrary elements.

This more general kind of measure on the basic figures of the first kind will be investigated more closely in the paragraphs below. One obtains as many essentially different measures as there are essentially different linear transformations of the basic figures. However, such transformations are of only two kinds:

1. Those that fix two (real or imaginary) elements of the basic figure (general case).

2. Those that fix only one (doubly counted) element of the basic figure (special case).

Correspondingly, there are also only two essentially different kinds of projective measure on the basic figures of the first kind: a *general* one, using transformations of the first kind, and a *special* one, using transformations of the second kind.

The ordinary measure on a line pencil is of the first kind, because rotation of the pencil about its centre fixes two distinct rays, namely those passing through the imaginary circular points at infinity.

On the other hand, the ordinary measure on the line is of the second kind. Under the assumption of ordinary parabolic geometry, displacement of a line along itself leaves only one point fixed, namely the point at infinity.

This already indicates how the assumption of hyperbolic or elliptic geometry causes the measure on the line to lose the special character conferred on it by parabolic geometry. Hyperbolic geometry gives the line two real points at infinity, and elliptic geometry gives it two imaginary points at infinity. The corresponding displacements of the line are linear transformations which fix two separate points at infinity. This will be discussed in more detail below.

---

[45]Here there is restriction on the kind of linear transformation that can be used. In the first place, the linear transformation must be real, so that it carries a real first element to a real second element. It is also necessary that scale elements be spaced in their order of construction, and we do not have, say, the first and second elements separated by the third and fourth. Cf. the text below.

§3
The general projective measure on the basic figures of the first kind

Initially we want to focus only on the general case of projective measure, where there are two fixed elements for the linear transformation that generates the scale. The latter may be called the *fundamental elements.* We use them to set up a coordinate system in which each other element is given by the ratio of two homogeneous variables $x_1 : x_2$. The value of this ratio will be denoted by z, so that $z = 0$ and $z = \infty$ represent the two fundamental elements. Then the linear transformation which initiates the construction of the scale is given by an equation of the form

$$z' = \lambda z,$$

where λ is a constant determining the transformation.[46] If we now apply this transformation repeatedly to an arbitrary element $z = z_1$, then we obtain the series

$$z_1, \quad \lambda z_1, \quad \lambda^2 z_1, \quad \lambda^3 z_1, \quad \ldots$$

and this series of elements is our scale. This series is mapped into itself by the generating transformation, as is clear *a priori.*

If we now designate the *scale interval as the unit of displacement,* then the distances of the elements $z_1, \lambda z_1, \lambda^2 z_1, \lambda^3 z_1, \ldots$ from the element z_1 are $0, 1, 2, 3, \ldots$ respectively.

To be able to measure the distances of other elements from z_1, we now turn to subdividing the scale interval, say into n (equal) parts. This is achieved by subjecting the end element of an interval to a transformation whose n^{th} iterate is the transformation $z' = \lambda z$, i.e. the transformation

$$z' = \lambda^{\frac{1}{n}} z.$$

The nth root here must be chosen so that the element $\lambda^{\frac{1}{n}} z$ lies between[47] the elements z and λz.

[46] As remarked above, this λ may not be completely arbitrary, since we have only real elements of the basic figure in view when constructing the scale. Thus λ must be such that $z' = \lambda z$ sends real elements to real elements (whether or not the fundamental elements $z = 0$, $z = \infty$ are real or imaginary). Also (see text below) λ must be positive for real fundamental elements.

[47] The best way to see why we make this particular determination is through the example of circle division. Giving the circle a scale interval, say one degree, is at first an indeterminate problem, because the given scale interval is determined only up to a multiple of the period 2π. This indeterminacy is removed by the above agreement. With real fundamental elements it suffices to define $\lambda^{\frac{1}{n}}$ simply as the positive real n^{th} root of λ. However, for this to exist, λ must be positive, as has already been stipulated. For negative λ one obtains a series of scale elements not arranged in order of distance.

When this subdivision is carried out one can define the distance from z_1 of all points with coordinates of the form

$$z = \lambda^{\alpha+\frac{\beta}{n}} z_1,$$

where α and β are integers. This distance is simply the exponent $\alpha + \frac{\beta}{n}$.

Now by allowing the subdivision of the scale to proceed without limit, it is clear that the distance from z_1 of any element z whatever should be regarded as the exponent α to which λ must be raised so that $\lambda^\alpha z_1 = z$. Here α is any rational or irrational number.

Since obviously $\alpha = \log \frac{z}{z_1} : \log \lambda$, we can also express this as follows:

The distance between an element z and the element z_1 is the logarithm of the quotient $\frac{z}{z_1}$, divided by the constant $\log \lambda$.

The element z_1 here is only accidentally chosen as the origin of the scale, and is not otherwise distinguished. One may move it arbitrarily by a linear transformation without changing the fundamental elements or, consequently, the measure. One therefore has

The distance between arbitrary elements z and z' is equal to

$$\log \frac{z}{z'} : \log \lambda,$$

as one may verify by taking the difference between the distances of the two elements z and z' from z_1, namely $\log \frac{z}{z_1} : \log \lambda$ and $\log \frac{z'}{z_1} : \log \lambda$.

Instead of the constant $\frac{1}{\log \lambda}$ we shall now write c for short,[48] a notation which will always be used below.

With this expression for the distance between two elements one easily verifies the existence of those properties the construction was intended to achieve. First there is the additivity of measure:

$$c \ \log \frac{z}{z''} = c \ \log \frac{z}{z'} + c \ \log \frac{z'}{z''}.$$

Also, the distance of an element from itself is zero:

$$c \ \log \frac{z}{z} = 0.$$

Finally, the distance between two elements,

$$c \ \log \frac{z}{z'}$$

[48] Corresponding to the restrictions on the constant λ, one has restrictions on the constant c. They are that c must be real or pure imaginary, according as the fundamental elements are real or imaginary. If c is chosen otherwise, one still has the above analytic expression as a measure, but the distance between two consecutive real elements is now imaginary.

is unaltered when z and z' are both subjected to a linear transformation fixing the fundamental elements

$$z = 0, \quad z = \infty,$$

that is, a transformation sending z and z' to the same multiple of themselves.

The analytic expression for distance obtained here admits a simple geometric interpretation. As is well known, the quotient $\frac{z}{z'}$ may be interpreted as the cross-ratio of the elements z, z' with the fundamental elements $z = 0$, $z = \infty$.

Therefore, our measure of the distance between two elements of the basic figure is a certain constant multiple of the logarithm of the cross-ratio of the given elements with the two fundamental elements.

The constant c in question is indeterminate and may be chosen arbitrarily.

§4
Passage to complex elements. Generalisation of the coordinate construction

Until now we have based the construction of a scale, and hence the definition of measure, only on two real elements of the basic figure. However, now that we have obtained the analytic expression for the distance between two elements,

$$c \log \frac{z}{z'},$$

we can also speak of the distance between two complex elements of the basic figure. Here is where we encounter a phenomenon, previously observed for angles, and soon to be seen for real elements whenever the fundamental elements are imaginary. It is that *the distance between two elements is not determined uniquely; it is a many-valued function with a modulus of periodicity.*

Since the logarithm function has period $2\pi i$, this modulus is $2c\pi i$.

Moreover, since the logarithm becomes infinite when its argument is 0 or ∞, elements for which $\frac{z}{z'} = 0$ or ∞ are obviously infinitely far apart. This happens if and only if one of the two elements is a fundamental element ($z = 0$ or $z = \infty$). Thus:

Under our measure the basic figure has two (real or imaginary) elements at infinity: the two fundamental elements.

In the same way, the distance between these two elements is infinite, being $\log 0$ or $\log \infty$.

The two fundamental elements are logarithmically infinitely far apart.

We now also drop the restrictive assumption we have so far made about the determination of coordinates. The two fundamental elements need no longer be taken as the basis elements of the coordinate system, but instead may be given by a general equation of second degree:

$$\Omega = az^2 + 2bz + c = 0,$$

or, in homogeneous notation,

$$ax_1^2 + 2bx_1x_2 + cx_2^2 = 0.$$

In order to give the distance between two elements with the homogeneous coordinates x_1, x_2 and y_1, y_2 one needs only to construct their cross-ratio with the two solutions of $\Omega = 0$. By well-known rules, the latter is

$$\frac{\Omega_{xy} + \sqrt{\Omega_{xy}^2 - \Omega_{xx}\Omega_{yy}}}{\Omega_{xy} - \sqrt{\Omega_{xy}^2 - \Omega_{xx}\Omega_{yy}}},$$

where Ω_{xx}, Ω_{yy}, Ω_{xy} denote the following expressions. Ω_{xx}, Ω_{yy} are the results of substituting x_1, x_2 resp. y_1, y_2 in Ω, thus:

$$\Omega_{xx} = ax_1^2 + 2bx_1x_2 + cx_2^2, \quad \Omega_{yy} = ay_1^2 + 2by_1y_2 + cy_2^2.$$

And Ω_{xy} denotes the expression

$$\Omega_{xy} = ax_1y_1 + b(x_1y_2 + x_2y_1) + cx_2y_2.$$

In this notation the distance between two elements is

$$c \log \frac{\Omega_{xy} + \sqrt{\Omega_{xy}^2 - \Omega_{xx}\Omega_{yy}}}{\Omega_{xy} - \sqrt{\Omega_{xy}^2 - \Omega_{xx}\Omega_{yy}}}$$

and this is the general analytic expression for distance.

Occasionally we will use the arc cosine in place of the logarithm. As is well known,

$$c \,\log a = 2ic \,\arccos \frac{a+1}{2\sqrt{a}},$$

and thus distance is also

$$2ic \,\arccos \frac{\Omega_{xy}}{\sqrt{\Omega_{xx}\Omega_{yy}}}.$$

The latter is the form of the analytic expression appearing in Cayley. As already mentioned, Cayley used only the special value $-\frac{i}{2}$ for c, so that for him the distance precisely equals the arc cosine in question.

§5
Special properties of the real elements of the basic figure

We shall now consider what the measure on basic figures of the first kind, constructed in the last two paragraphs, says in particular about the real elements of the figure. Here we have to distinguish two cases, according as the fundamental elements are real or imaginary. To fix ideas we shall concentrate on the measure on the linear point series; obviously the same applies to the line pencil.

Suppose *first* that two real fundamental points o, o' are given on the line.

Then if x and y are real points of the line, x, y have a negative or positive cross-ratio with o, o' according as the segment xy is separated by the segment oo' or not. Thus in the first case the logarithm of the cross-ratio is pure imaginary, in the second it is real (up to an imaginary period). Hence if we make the requirement that the distance between two neighbouring points on the line be real, we must take the constant c multiplying the logarithm to be real. Then we have the theorem:

The distance between two points x, y is imaginary or real according as the segment xy is separated by the segment oo' or not.

Of course, one could give c a pure imaginary value (as Cayley did); then the words "real" and "imaginary" would have to be exchanged in the preceding theorem. At the outset this is just as admissible as the other assumptions. But then the measure, considered for real points, would have a character radically different from what we are used to. E.g. if we wished to construct a scale of such points $1, 2, 3, \ldots$ unit distances apart, then 2 would have to be separated from 1 and 3 by oo', and the distance 13 would only be two units when measured first from 1 to 2, then from 2 to 3; direct measurement of $\overline{13}$ would give an imaginary value, etc. For this reason, the assumption of an imaginary c will be excluded here.

For real c we have, first of all, the theorem just given. Accordingly, we confine ourselves to considering one of the two segments into which the line is divided by the two fundamental points. Each of these segments is infinitely long, inasmuch as its ends, the fundamental points, are infinitely far from all other points.

Now imagine being placed at a point on the segment oo' just considered, and being able to move only according to those linear transformations that leave the points o, o', and hence the measure, unaltered. We can then speak of the speed of motion, as the ratio of the distance covered to the time taken. If one moves with constant speed in either direction along the line then one constantly approaches the point o or o' but, since it is infinitely distant, never reaches it. Moreover, *one never reaches the second segment $o'o$, so one would not be able to convince oneself of its existence.*

Now this is just the idea one has in *hyperbolic* geometry of measuring along a line. Hyperbolic geometry provides each line with two points at infinity. Whether there is a piece of the line beyond the points at infinity, completing the piece lying in the finite to a closed curve, we cannot say, since we cannot move as far as the points at infinity, let alone beyond them. In any case, one can add such a piece as an imaginary or ideal segment of the straight line.

We now wish to make a *second* assumption, that the two fundamental points used to determine the measure on the line are (conjugate) imaginaries. Then the cross-ratio of the two fundamental points to any two real points x, y has absolute value 1, so its logarithm is pure imaginary. We must therefore give c a pure imaginary value $c_1 i$ in order to have a real distance between real points. But then the mutual distance between real points is also real. There are no real points at infinity. The line turns around like a closed curve. The real distance between two points is not completely determined, only up to a multiple of a real period representing the total length of the line. The latter has the value $2i\pi c = -2\pi c_1$. The measure on the line is therefore exactly like the ordinary measure on a circle of radius c_1.

The measure just described on the line is precisely that belonging to *elliptic* geometry.

What we have just done for the linear point series can be done similarly for the *line pencil.*

If the two fundamental rays used to determine measure in the pencil are real, then the pencil has two real rays which make an infinitely large angle with all other rays. Rotation of a ray in the pencil – defined like the motion of a point along the line above – never carries the ray as far as either of the limit rays, or beyond them. Such a measure is certainly no basis for our usual determination of angle, since continued rotation of a ray will return it to its initial position after a finite amount of time. Rather, this property requires imaginary fundamental rays. And in fact we already noticed in §2 that ordinary angle measure uses two imaginary fundamental rays, namely the two rays of the pencil that go through the imaginary points at infinity on the circle. In hyperbolic and elliptic geometry the angle measure in a pencil remains much the same as usual; only the fundamental rays are no longer defined as the rays through the two points of the circle, but as those touching a particular conic section, the circle at infinity for this geometry (cf. §8).

The constant c left undetermined in the general formula of §4 is here set equal to $\pm\frac{\sqrt{-1}}{2}$, as for ordinary angle measure. In the first place, it must be a pure imaginary $\pm c_1 i$ for the reasons explained in connection with the linear point series. Then the total angle of the pencil becomes $2\pi c_1$, and since the usual value[49] for this is π, we have to take $c_1 = \frac{1}{2}$. Under this assumption,

[49]The total angle of a pencil is here understood as the angle through which a line must

the formula in §4 becomes the usual formula for the determination of angles. Let x and y be rectangular coordinates in the plane. We consider a pencil with centre at the origin. Then the two rays through the points of the circle at infinity are

$$x^2 + y^2 = 0.$$

Now let two rays be given by homogeneous coordinates x, y and x', y'. Then if we set $c = \frac{i}{2}$ in the formula of §4, the angle between them becomes

$$\arccos \frac{xx' + yy'}{\sqrt{x^2 + y^2}\sqrt{x'^2 + y'^2}},$$

and this is clearly the usual value.

Thus the angle between two lines is defined in projective geometry as $\frac{\sqrt{-1}}{2}$ *times the logarithm of the cross-ratio of the two lines to the lines from their common point to the imaginary points at infinity of the circle.*

In particular, straight lines meet at a right angle when their cross-ratio is harmonic. The identification of such an angle (and also a corresponding segment) as a right angle will also be used occasionally with general measures below.

§6
The special measure with coincident fundamental elements

Until now we have not considered the special case where there is coincidence of the two fundamental elements used to define the measure. Our general formula

$$2ic \arccos \frac{\Omega_{xy}}{\sqrt{\Omega_{xx}\Omega_{yy}}}$$

then gives a distance of zero between the points represented by x and y, independently of the values of these coordinates. However, construction of a measure is still possible, because of the quite definite way the distance between different elements tends to zero as the fundamental elements tend to coincide. We obviously have

$$\Omega_{xx}\Omega_{yy} - \Omega_{xy}^2 = (x_1y_2 - y_1x_2)^2\Delta,$$

where Δ is the discriminant $(ac - b^2)$ of the quadratic form Ω. It follows that we can also write the general formula for distance as

$$2ic \arcsin \frac{(x_1y_2 - y_1x_2)\sqrt{D}}{\sqrt{\Omega_{xx}\Omega_{yy}}}.$$

turn in order to coincide with its initial position. This is half the angle a point on the circle must traverse in order to return to its initial position.

Now if the two fundamental elements tend to coincide, Ω becomes a perfect square of a linear expression $p_1x_1 + p_2x_2 = p_x$ and Δ vanishes. We can therefore replace the arcsine by the sine itself, and hence write the distance as:

$$2ic\sqrt{\Delta}\frac{x_1y_2 - y_1x_2}{\sqrt{\Omega_{xx}\Omega_{yy}}},$$

or, substituting p_x^2 and p_y^2 for Ω_{xx} and Ω_{yy}:

$$2ic\sqrt{D}\frac{x_1y_2 - y_1x_2}{(p_1x_1 + p_2x_2)(p_1y_1 + p_2y_2)}.$$

We combine the vanishing factor $\sqrt{D}$ with the $2ic$, to which we can assign an arbitrarily large value, obtaining a new constant k. *In this way we obtain the following formula for the difference of measures:*[50]

$$k\frac{x_1y_2 - y_1x_2}{(p_1x_1 + p_2x_2)(p_1y_1 + p_2y_2)},$$

where $p = 0$ represents the doubly-counted fundamental element.

It is easy to verify that this expression, which we have found by a limit process, in fact satisfies the conditions laid down in §2.

For this purpose we write it in the somewhat different form

$$\frac{q_1x_1 + q_2x_2}{p_1x_1 + p_2x_2} - \frac{q_1y_1 + q_2y_2}{p_1y_1 + p_2y_2}$$

where q_1, q_2 are otherwise arbitrary quantities satisfying the condition

$$q_1p_2 - p_1q_2 = k.$$

This form shows first and foremost the additivity of the measure.

But we also get its invariance under the special linear transformations that fix the doubly counted element $p = 0$. These transformations carry p to a multiple of itself. Every other linear expression, and hence also q, goes to the same multiple of itself, plus a multiple of p:

$$p' = \rho p,$$
$$q' = \rho q + \sigma p.$$

The quotient $\frac{q}{p}$ is thereby altered by the constant $\frac{\sigma}{\rho}$, and the distance between two elements, which is the difference of two such quotients, does not alter at all, q.e.d.

The measure found here may be defined geometrically as follows. As is well known, the quotient $\frac{p_x}{q_x}$ represents the cross-ratio of the point x and the

[50] Cayley derives this formula in much the same way.

point for which $\frac{q}{p}$ has the value 1 with the two points $p = 0$, $q = 0$, and hence it is the cross-ratio of the given element with an arbitrarily chosen element and the given doubly-counted fundamental element. *The difference between the values of this cross-ratio, for two given elements, represents the distance between them.*

This measure, which arises as a limiting case of the general one, will be called the *special measure.* In contrast to the general measure it has the following two special properties.

It is no longer many-valued, but single-valued.

Instead of having two logarithmically infinitely distant elements, it has just one, algebraically infinitely distant element (the doubly-counted fundamental element).

This special measure subsumes the ordinary (euclidean, parabolic) measure on the line, as already indicated in §2. This is why the line, as normally intuited, has only one point at infinity. The point at infinity can be approached from either side, without ever being reached. In parabolic geometry, in contrast to elliptic geometry, the straight line is infinitely long. But it no longer possesses an ideal segment, like the line in hyperbolic geometry; it closes up at infinity.

§7
Special measures tangential to a general measure at a particular element. Curvature of the latter

We now consider two measures on a basic figure of the first kind, one general and one special. They will be related in a certain way we shall call *tangency of the measures at an element.* The nature of this relationship is best explained by an example.

Consider the ordinary measure on a line, using the point at infinity as the double element. Let the points of the line be given by a non-homogeneous coordinate z, which is just the distance from the origin.

Now construct a general measure on the line as follows. At distance 1 from the given line, on the perpendicular through the origin, take the centre of a pencil of lines. On this pencil again take the ordinary measure, i.e. the ordinary angle measure. The latter measure can be carried over to the given line by defining the distance between two points to be the angle between the lines in the pencil that go through them. If z is the coordinate of a point, then the angle between the line through it and the line through the origin is $\arctan z$. Thus, with this measure, the distance between the two elements z and z' is

$$\arctan z - \arctan z'.$$

The fundamental elements for this measure are imaginary and correspond to $z = \pm i$.

The relationship between the special measure and the general one just constructed is that for values of z near $z = 0$ they nearly coincide, since for small angles the differences between the angle and its tangent vanishes.

This is seen most clearly when we replace $\arctan z$ by its series expansion $z - \frac{z^3}{3} + \frac{z^5}{5} - \cdots$. *Thus in the neighbourhood of* $z = 0$ *the two measures agree up to terms of third order.* This relation between the two measures is what we call *tangency.*

When a general and a special measure are tangential, as above, then it is obvious that the point of tangency is the fourth harmonic point of the two fundamental points of the general measure and the doubly-counted fundamental point of the special measure. Thus if one wants to construct a special measure on a basic figure, tangential to a given general measure at a particular element of the figure, one first finds the fourth harmonic of this particular element with the two fundamental elements of the general measure. The latter is the double element from which the desired measure is constructed. Then it only remains to choose the absolute value of the constant in this measure so as to obtain agreement between the two measures in the neighbourhood of the given element. Because of the particular position of the doubly-counted fundamental element, this agreement is necessarily intimate, a tangency.

There is a characteristic difference between the general measure with imaginary fundamental elements and the general measure with real fundamental elements.

If the two fundamental elements are imaginary, then the special measure runs ahead of the general measure near the point of tangency. In the example just given, that means that the distance of a point z from the origin, measured by the tangential special measure, is always greater than the distance between the same points measured by the given general measure. This is clear when one reflects that the whole line, according to the special measure, has infinite length, whereas it has finite length in the general measure. The two measures agree only for points infinitely close to the point of tangency ($z = 0$).

On the other hand, if the two fundamental elements for the given general measure are real, then the special measure stays behind the general measure near the point of tangency. In fact, the segment between the two fundamental elements is already infinitely large in the general measure, but only finite for the tangential measure.

This staying behind or running ahead of the general measure relative to the special measure is what I call *curvature of the general measure*, initially at the point of tangency. The curvature will be called *positive* when the fundamental elements are imaginary, and *negative* when they are real. The *measure of curvature*, roughly speaking, is the amount of staying back or running ahead. As I shall show, the curvature has the same value for all elements of the figure. It can be taken to be $-\frac{1}{4c^2}$, where c is the characteristic constant for the general measure.

The two fundamental elements of the general measure may be taken harmonic to $z = 0$ and $z = \infty$, as in the above example, and more specifically may be determined by the equation

$$z^2 = a.$$

Then if c, as always, is the characteristic constant of the measure, one finds the distance of an element z from the origin to be

$$2c \arctan \frac{zi}{\sqrt{a}}$$

or, expanded in series,

$$\frac{2ci}{\sqrt{a}} z + \frac{2ci}{3\sqrt{a^3}} z^3 + \frac{2ci}{5\sqrt{a^5}} z^5 + \cdots .$$

The tangential special measure at the element $z = 0$ is that which uses the first term of the expansion as the distance of the element z from 0, namely the expression

$$\frac{2ci}{\sqrt{a}} z.$$

One can then define the deviation of the general measure from its tangential special measure, or the curvature, to be the negative of the second term of the expansion, multiplied by 3, and divided by the cube of the first term. The result is the given expression $-\frac{1}{4c^2}$.

This expression for the curvature also has the sign settled on above. With real fundamental elements one must (§4) give c a positive sign, thus the curvature is negative. With imaginary fundamental elements one must give c a pure imaginary value, so the measure of curvature will be positive. The curvature of a special measure is zero, because in the passage to the special measure in the previous paragraph we had to give c an infinite value.

Finally, the curvature is the same at all elements of our basic figure, inasmuch as c has the same meaning for all elements.

The curvature of a general measure can also be defined in the following way.

In §5 it was shown that the length of the whole line with imaginary fundamental elements and constant $c_1 i$ was equal to $2c_1\pi$. The reciprocal of the square of this expression, multiplied by π^2, is just the curvature. The curvature is π times the area of the ordinary circle with radius equal to the reciprocal of the apparent length of the whole line.

With real fundamental elements one can view the situation as follows. The distance between the two fundamental elements $z = \pm\sqrt{a}$, measured by the tangential special measure, is $4c$. Thus the curvature equals -4 times the square of the reciprocal of the special measure distance between the two fundamental elements.

We remark in conclusion that the three kinds of measure, giving elliptic, hyperbolic and parabolic geometry on the line, are tangentially related to each other. The tangency occurs at whichever point we take as the origin of measurement in the sense of hyperbolic, elliptic or parabolic geometry. The measure of parabolic geometry is the special measure tangential to the general measure of elliptic of hyperbolic geometry. For this reason it can replace the latter for all points sufficiently near the origin.

§8
The general projective measure in the plane

Having explained projective measure on the basic figures of the first kind, we can now proceed almost immediately to projective measure on the basic figures of second and then arbitrary dimension. We find a more general measure which includes as special cases the measures we normally use on these basic figures, together with measures established by elliptic and hyperbolic geometry on the figures in question. These will first be explained for the plane. The situation is quite similar for points (as well as for pencils of lines and planes in space), as will be discussed further in §10.

Just as one uses two fundamental elements on a basic figure of the first kind to determine a projective measure, a projective measure on the plane is based on conic section called the *fundamental conic* (called by Cayley "the absolute"). This fundamental conic determines a measure on all basic figures of the first kind in the plane, i.e. on all lines and line pencils in the plane. Each line cuts the fundamental conic in two points (real, imaginary or coincident). These will be the fundamental points for the measure on it. Among the lines in a pencil there are two tangents to the conic (real, imaginary or coincident). These will be the fundamental rays for the measure on the pencil. It then remains to make a choice of the constants c. We shall use the same, arbitrarily chosen, constant c for all linear point series; similarly, to construct the measure on all line pencils we shall use the same, arbitrarily chosen, constant c'.

Now we shall present the analytic expression for the measure just introduced.

Suppose that the equation for the fundamental conic, in point coordinates, is

$$\Omega = 0.$$

Let

$$\Omega_{xx}, \quad \Omega_{yy}$$

denote the expressions obtained by substituting in Ω the coordinates x_1, x_2, x_3 of a point (x), and the coordinates y_1, y_2, y_3 of a point (y), respectively. Finally, let

$$\Omega_{xy}$$

denote the result of substituting the coordinates y in the polar of (x), or what comes to the same thing, the coordinates x in the polar of (y). Then the cross-ratio of the two points (x) and (y) to the two points where their connecting line meets the fundamental conic is given by the quotient of the roots of the following quadratic equation for λ:

$$\lambda^2\Omega_{xx} + 2\lambda\Omega_{xy} + \Omega_{yy} = 0.$$

The cross-ratio is therefore equal to

$$\frac{\Omega_{xy} + \sqrt{\Omega_{xy}^2 - \Omega_{xx}\Omega_{yy}}}{\Omega_{xy} - \sqrt{\Omega_{xy}^2 - \Omega_{xx}\Omega_{yy}}},$$

and the *distance between the two points* is

$$c \log \frac{\Omega_{xy} + \sqrt{\Omega_{xy}^2 - \Omega_{xx}\Omega_{yy}}}{\Omega_{xy} - \sqrt{\Omega_{xy}^2 - \Omega_{xx}\Omega_{yy}}}$$

or, the same thing,

$$2ic \arccos \frac{\Omega_{xy}}{\sqrt{\Omega_{xx}\Omega_{yy}}}.$$

Thus with the notation used here, we get exactly the same expressions as for the basic figures of the first kind.

In much the same way we obtain the angle between two lines with the coordinates u_1, u_2, u_3 and v_1, v_2, v_3 when

$$\Phi = 0$$

is the equation of the fundamental conic in line coordinates. The *angle between the two lines is*

$$c' \log \frac{\Phi_{uv} + \sqrt{\Phi_{uv}^2 - \Phi_{uu}\Phi_{vv}}}{\Phi_{uv} - \sqrt{\Phi_{uv}^2 - \Phi_{uu}\Phi_{vv}}},$$

or, the same thing,

$$2ic' \arccos \frac{\Phi_{uv}}{\sqrt{\Phi_{uu}\Phi_{vv}}}.$$

The first question that now arises is: where are the points (y) equidistant from a point (x)? Since the distance $\overline{xy}$ now depends only on

$$\frac{\Omega_{xy}}{\sqrt{\Omega_{xx}}\sqrt{\Omega_{yy}}},$$

we obtain the equation of the locus of (y) by setting this expression equal to a constant k or, what comes to the same thing, setting

$$\Omega_{xy}^2 = k^2 \Omega_{xx} \Omega_{yy}.$$

But this is a conic section which touches the fundamental conic $\Omega_{yy} = 0$ at its two intersections with $\Omega_{xy} = 0$, the polar of (x) with respect to the fundamental conic.

The points on a conic touching the fundamental conic at its intersections with the polar of the point (x) *are equidistant from the point* (x).

These conics of equidistant points are what we would ordinarily call *circles.* The point (x) is the common *centre* of the circles. The *radius* of a circle is the distance of any of its points (y) from the centre (x), i.e. the expression

$$2ic \arccos k.$$

Among the circles with centre x there is, in particular, $k = 0$, and thus a radius of πic, corresponding to the polar of the point (x). They also include (for $k = 1$) a circle of radius zero. The latter consists of the pair of tangents to the fundamental conic through (x). The distance between two points on such a tangent is in fact always zero, because they have cross-ratio $+1$ with the intersections of their connecting line with the fundamental conic. One must give an infinite value to the constant c, which will not do here, since it makes the distance infinite between any points not on a tangent to the fundamental conic. Of course, we are at liberty to give c an infinite value just for the measure on tangents to the fundamental conic. However, this measure is no longer comparable with the general measure. Finally, among the concentric circles in question there is one with an infinite radius, corresponding to $k = \infty$. This is the fundamental conic itself, because the points where the line through (x) meets the fundamental conic are both infinitely distant: *the fundamental conic is the locus of points at an infinite distance from a given point.*

Quite similar considerations as those for points in the plane can also be applied to lines.

Those lines that make the same angle with a fixed line (u) *envelope a conic which touches the fundamental conic at its two intersections with* (u)*; this includes in particular the pole of* (u) *(regarded as a pencil of lines). Those lines that go through one of the intersections of the fundamental conic* *with (u) make a zero angle with (u). The tangents to the fundamental conic make an infinite angle with (u).*

Thus those conics that touch the fundamental conic twice are the loci of those points that have a constant distance from a fixed point, the pole of the chord joining the contact points. And they are enveloped by those lines making a constant angle with a fixed line, the chord just mentioned. Also note the following. One calls lines *parallel* if they meet at infinity, i.e. on the fundamental conic. The angle between two parallel lines is zero. However, there is nothing to prevent the introduction of a *special* measure on a pencil of parallel lines, by taking an infinitely large value of c. Such a special measure comes to light when we speak of a distance[51] between two parallels in ordinary parabolic geometry.

§9
The linear transformations of the plane that play the rôle of motions

A conic section has the property of being transformed into itself by a triply-infinite family of linear transformations of the plane. This is because there is an eightfold infinity of linear transformations of the plane and only a fivefold infinity of conic sections, so that each conic is mapped to any other, and to itself in particular, by a threefold infinity of linear transformations.

Such a linear transformation of the plane permutes the points of the conic section among themselves, exactly like a linear transformation of a basic figure of the first kind. One concludes from this that each such linear transformation leaves two points of the conic section fixed. Indeed, consider the two line pencils $o(p_1, p_2, p_3, \dots)$ and $o(p'_1, p'_2, p'_3, \dots)$ going from a fixed point o of the conic to any of its other points $p_1, p_2, p_3, \dots,$ and to their images $p'_1, p'_2, p'_3, \dots$ under a linear transformation of the plane that maps the conic onto itself. The two pencils are projective, because $o(p_1, p_2, p_3, \dots)$ is projective with $o'(p'_1, p'_2, p'_3, \dots)$ where o' denotes the image of o under the transformation. But this point o', like o, is a point of the given conic section, so $o'(p'_1, p'_2, p'_3, \dots)$ is projective with $o(p'_1, p'_2, p'_3, \dots)$ and thus the latter is also projective with $o(p_1, p_2, p_3, \dots)$, q.e.d. But if the two pencils $o(p_1, p_2, p_3, \dots)$ and $o(p'_1, p'_2, p'_3, \dots)$ are projective, then they have two rays $o\pi_1$ and $o\pi_2$ in common, and hence there are two points π_1, π_2 of the conic unaltered by transformation.

Now if two points of the fundamental conic are unaltered, so are the line connecting them, the tangents to the conic at these points and their point of intersection, and hence the triangle formed by the connecting line and the

[51]It is a peculiarity of parabolic geometry that the distance between two parallels equals the minimal distance between points on them.

two tangents. By taking this triangle as coordinate triangle, the equation of the conic becomes:

$$x_1x_2 - x_3^2 = 0.$$

Since the linear transformation mapping the conic onto itself must leave the triangle unaltered, it is of the form:

$$x_1 = \alpha_1 y_1, \quad x_2 = \alpha_2 y_2, \quad x_3 = \alpha_3 y_3.$$

The condition for this transformation to preserve the conic is

$$\alpha_1\alpha_2 - \alpha_3^2 = 0,$$

and since this is only a condition on the three homogeneous α, there is a simple infinity of linear transformations that preserve both the triangle and the conic.

Under these transformations the quotient $\frac{x_1x_2}{x_3^2}$ remains unaltered, independently of its value. Thus the transformations in question map all conics of the form

$$x_1x_2 - kx_3^2 = 0$$

onto themselves.[52]

Here we must distinguish between real conics with real points and real conics without real points.

In the former case the linear transformations mapping the conic into itself, as long as they are real, fall into two families. *The transformations in the first family can be generated by repetition of a real infinitesimal transformation of the same kind and therefore constitute a group; those of the second family do not.*

For example, if the two fixed points on the conic are π_1 and π_2, a transformation belongs to the first or second family according as $\alpha_3 = \sqrt{\alpha_1\alpha_2}$ or $\alpha_3 = -\sqrt{\alpha_1\alpha_2}$. In the latter case the segment $\pi_1\pi_2$ is separated by any two corresponding points on the conic; in the former case it is not.

In the case under consideration, the transformations in the first family are those we shall call *motions* of the plane. For it is these which become real motions when we particularise the fundamental conic in a suitable way, so that the measure derived from it becomes the one actually used.[53]

---

[52] Incidentally, one sees from this that not every linear transformation maps a conic section into itself; however, if the transformation does this for *one* conic section, it does so for infinitely many.

[53] The other family of transformations of the conic become, at the same time, those transformations of the plane which send plane figures to their inverses.

If on the other hand the basic conic is given by a real equation but has no real points, then all real linear transformations mapping the conic into itself will be called motions.[54]

With this definition, the theorem just proved that each linear transformation preserving the given conic also preserves infinitely many others, can be expressed as follows.

A motion of the plane not only preserves the fundamental conic section, but also each conic (each circle) touching it at the two fixed points.

These conics include the point $x_1 = 0, x_2 = 0$ which is the common centre of the circles. We shall call the motion a *rotation of the plane about this centre.*

Then we have the theorem:

Each motion of the plane is a rotation about a point. All other points describe circles with this point as centre.

It scarcely needs to be stressed that, under a motion, the polar of the centre of rotation dualistically plays the same rôle as the centre, and hence that motion is a self-dual concept under our measure. This duality is only annulled when we undualistically particularise the fundamental conic to obtain parabolic geometry.

Among the motions of the plane there is a special case, which occurs when the centre of rotation is at infinity, i.e. on the fundamental conic.

The circles described by the points of the plane are then conic sections having quadruple contact with the fundamental conic at the centre. The kind of motion occurring under this assumption will be called a *translation.*[55]

It is now clear that, when motions are defined in this way, we have the theorem:

Metric relations are unaltered by motions of the plane.

This is because a motion maps the fundamental conic into itself, and hence preserves the cross-ratio of two points with the intersections of their connecting line and the fundamental conic. Hence the same is true of a constant multiple of the logarithm of the cross-ratio, i.e. of the distance between the two points. Similarly for the angle between two lines.

This is true not only for motions of the plane but also, on the same grounds, for transformations of the second kind mapping the fundamental conic into itself [whenever this class is distinct].

Moreover, something similar holds for the dual transformations mapping the fundamental conic into itself, in particular for the polar reciprocity de-

---

[54] [The distinction between direct and inverse congruence disappears in this case because the complete projective plane, and hence also the elliptic plane, is a one-sided surface. See the discussion in the 2nd volume of these works.](Translator's note. Klein's *Gesammelte Mathematische Abhandlungen.*)

[55] A peculiarity of the particularisation of the fundamental conic is that in parabolic geometry the translations form a closed system, and any two of them commute.

termined by this conic. In fact, two points and the intersections of their connecting line with the fundamental conic, having a certain cross-ratio, correspond under these transformations, to two lines and tangents to the fundamental conic through their intersection, which have the same cross-ratio. Thus if we take the constants c and c' (§8) for the two measures to be equal, we have the theorem:

The distance between two points equals the angle between the corresponding lines, and conversely;

in particular:

The distance between two points equals the angle between their polars.

We shall not make further use of these theorems here, and will only come back to the latter in the next paragraph. It subsumes the theorem from spherical geometry that the sides and angles of a spherical triangle are exchanged by passing to the polar triangle.[56]

§10
The general projective measure in pencils of lines and planes

In quite a similar way to the setting up of a general projective measure on the plane in the previous two paragraphs, we can set one up for the other basic figure of the second kind, the point (regarded as a pencil of planes or lines). In this case one uses a *fundamental cone of second degree* instead of the fundamental conic. The angle between two lines which meet at the apex of the cone is the constant c times the logarithm of the cross-ratio made by the two lines with the generators of the cone in the same plane. The angle between two planes which meet at the apex is another constant c' times the logarithm of the cross-ratio of the two planes with the two tangent planes to the cone through their line of intersection.

The analytic expression for this measure is exactly the same as that for the measure in the plane constructed above. One has only to give the coordinates (x), (y) and (u), (v) in the plane the meaning of line and plane coordinates at the point. Also, all other developments for the plane carry over immediately to the point; this hint should suffice here.

It is easy to see that the ordinary measure for a point,[57] i.e. the ordinary way of measuring the angle between lines or planes going through the point, is a special case of the general measure. *This measure takes the fundamental*

[56] Cf. Cayley, l.c.

[57] One does not ordinarily speak of the measure for a point, but for a sphere (of radius 1) centred on it. In the text the former terminology is preferred, since the point is the figure on which projective geometry operates. This is not to ignore the distinction, which has already appeared, between measure on a pencil of lines and measure on the circle. Each line going through the centre of the pencil corresponds to *two* points on the sphere (or circle). This leads to a different measure on the sphere (or circle) which would only be an unnecessary complication here.

cone of second degree to be the cone connecting the point to the imaginary circle at infinity,[58] *setting both the constants c and c' equal to $\frac{\sqrt{-1}}{2}$.*[59]

This is because, in a rectangular coordinate system, the cone of lines from the point to the imaginary circle at infinity has equation

$$x^2 + y^2 + z^2 = 0,$$

or, in plane coordinates:

$$u^2 + v^2 + w^2 = 0.$$

The angle between the lines with coordinates (x, y, z), (x', y', z'), and between the planes (u, v, w), (u', v', w'), is therefore obtained from the formulae of §8 by setting $c = c' = \frac{\sqrt{-2}}{2}$, namely

$$\arccos \frac{xx' + yy' + zz'}{\sqrt{x^2 + y^2 + z^2}\sqrt{x'^2 + y'^2 + z'^2}}$$

and

$$\arccos \frac{uu' + vv' + ww'}{\sqrt{u^2 + v^2 + w^2}\sqrt{u'^2 + v'^2 + w'^2}},$$

and this is the ordinary angle measure. The polar with respect to the fundamental conic, of a plane going through the point, is the perpendicular to it. The last theorem of the previous paragraph now becomes the following: the angle between two planes equals the angle between their normals. This theorem is the basis of the principle used in spherical geometry, that the measures of a spherical triangle and its polar triangle are dually related, i.e. sides and angles are exchanged.

§11
The measure on the plane with an imaginary conic. Elliptic geometry

The ordinary measure for a point is a propotype of the projective measure for a point or plane when the fundamental cone resp. fundamental conic is imaginary. The only particularisation occurring with the ordinary measure is that the two constants c and c' equal $\frac{\sqrt{-1}}{2}$. If they had the more general values $c_1\sqrt{-1}$ and $c'_1\sqrt{-1}$, the measures would differ only by factors $2c_1$, $2c'_1$:

$$2c_1 \arccos \frac{xx' + yy' + zz'}{\sqrt{x^2 + y^2 + z^2}\sqrt{x'^2 + y'^2 + z'^2}}$$

[58] In elliptic and hyperbolic geometry this must be replaced by the cone of tangents from the point to the second degree surface at infinity.

[59] This is the same choice of constants always made by Cayley.

and

$$2c'_1 \arccos \frac{uu' + vv' + ww'}{\sqrt{u^2 + v^2 + w^2}\sqrt{u'^2 + v'^2 + w'^2}},$$

expressions which can be derived by the same calculations as the originals.

Thus if an imaginary conic section is given in the plane then the length of each real line is finite, and so is the angle sum of a line pencil. If we retain the notations c_1 and c'_1 for the constants c and c'[60] divided by i, then the length of the line is $2c_1\pi$ and the angle sum of a pencil is $2c'_1\pi$.

There are neither real points at infinity, nor real lines that form infinitely large angles with others. It follows that all relations between angles of lines and planes through the same point convert to distances between points and angles between lines in the plane simply by dividing distance by $2c_1$ and angle by $2c'_1$. *Thus plane trigonometry, under this measure, is the same as spherical trigonometry*, except that sides and angles have to be divided by $2c_1$ and $2c'_1$ respectively.

The plane measure just described is precisely that for *elliptic* geometry. One specialises it further by setting the constant c'_1 equal to $\frac{1}{2}$, so that the angle sum of a pencil equals π, as with the usual measure for points. The angle sum of a plane triangle is then greater than π, as for spherical triangles, and only equal to π for infinitely small triangles, etc.

Thus one has a model for the planimetric part of elliptic geometry when one takes any imaginary conic section in the plane and uses it to construct a projective measure. For example, one could take the conic section to be the intersection of the plane with the cone from a point in space to the imaginary circle at infinity. One then sets c and c' equal to $\frac{\sqrt{-1}}{2}$. The distance between two points or the angle between two lines of the plane is equal to the angle subtended by the two points or the two lines at the chosen point in space. On the other hand, if the geometry given to us is indeed elliptic, then the infinitely distant points of the plane constitute an imaginary conic section, and the elliptic geometry coincides with the geometry of the projective measure constructed from this conic section.

§12
The measure on the plane with a real fundamental conic section. Hyperbolic geometry

We now suppose we are given a real fundamental conic section in the plane. This will lead to a measure for which the points inside the fundamental conic have the properties of hyperbolic geometry.

[60] c and c' are in fact pure imaginary, for the same reason that the constant c in §5 is imaginary for imaginary fundamental elements.

If the fundamental conic is real, the real points and lines of the plane fall into two classes. There are points which admit two real tangents to the conic, and points which admit none. The former are said to be outside the conic, and the latter inside. Similarly, the lines are divided into two groups: those which meet the conic in two real points, and those which meet it in two imaginary points.

The connection with hyperbolic geometry is found by restricting attention to the points inside the conic section, and the lines passing through them.

No line pencil with centre in the space considered by us has real elements at infinity. For this reason the constant c' is pure imaginary, and we take it to be $c'_1 i$. The angle sum of a pencil with centre inside the fundamental conic section is then $2c'_1\pi$.

On the other hand, each line passing through the region in question has two real points (logarithmically) at infinity, its intersections with the fundamental conic section. We therefore give a real value to the constant c.

With this assignment of constants c and c', all points inside the conic are a real distance apart. Similarly, all lines which meet inside the conic do so at a real angle. But the distance between two points separated by the fundamental conic is imaginary. The fundamental conic is the locus of the infinitely distant points. Two lines which pass inside the fundamental conic, but meet outside it, meet at an imaginary angle. Between them and the lines that meet inside are those transitional lines that meet on the fundamental conic, and thus at infinity – the lines called parallel in §8. Their angle is zero.

We now suppose that we are at any place inside the fundamental conic, and that we can move by means of those linear transformations that preserve the fundamental conic (cf. §5, §9). Then we shall be able to rotate, as under the ordinary measure, and return to our starting position after a finite amount of rotation. Likewise, we shall be able to move arbitrarily far in either direction along a straight line. *But we shall never reach the fundamental conic, let alone pass beyond it.* Thus we are trapped inside the conic; it is the limit of our plane; we cannot tell whether or not it is the whole plane. An observer, supplied with the ordinary measure, who saw us moving towards the fundamental conic at constant speed in our measure, would notice us moving more and more slowly (after a certain point) and failing to reach it.

The geometry based on this measure *corresponds completely with the idea of hyperbolic geometry*, when we set the so far undetermined constant c'_1 equal to $\frac{1}{2}$, making the angle sum of a line pencil equal to π. In order to be convinced of this, we consider a few propositions of hyperbolic geometry in somewhat more detail (the propositions themselves will be enclosed by quotes).

"Through a point in the plane there are two parallels to a given line, i.e. lines meeting the given line at infinity". These are the lines connecting the point to the intersections of the given line with the fundamental conic.

"The angle between the two parallels to a given line through a given point increases with the distance of the point from the line. As the point tends to infinity, this angle tends to π, i.e. the other angle between the two parallels tends to zero". In fact, when the point is on the fundamental conic the two parallels, like any two lines meeting at the fundamental conic, enclose a zero angle. Therefore, we also have the theorem: "The angle between a line and each of its parallels is zero". For points not at infinity, the "angle of parallelism" of hyperbolic geometry also occurs in the geometry of our projective measure, and it will in fact be shown that the trigonometric formulae of the two geometries coincide.

"The angle sum of a triangle is less than π; for a triangle with vertices at infinity the angle sum is zero". The latter follows from the fact that such vertices necessarily lie on the fundamental conic, and any two lines that meet there enclose a zero angle. The validity of the general theorem, which is made probable by the fact that infinitely large triangles have angle sum zero and infinitely small ones have angle sum π, will follow from the trigonometric formulae to be given below.

"Two perpendiculars to the same line do not meet". For us of course they do meet, at the pole of the line. However, the pole lies in the space outside the conic, the existence of which cannot be known through our motions. However, we can adjoin such a space – and this also happens in hyperbolic geometry – as an *ideal* space,[61] just as one adjoins an improper line at infinity to the real elements of the plane in parabolic geometry. Nothing is claimed about the existence of the ideal part of space; we use the expression only as a non-contradictory and convenient term.

"A circle of infinite radius is different from a line". A circle of infinite radius means, for us, a conic with four-point contact with the fundamental conic. On the other hand, a line, i.e. a line segment through the inside of the conic, is a circle whose centre (the pole of the line) lies in the ideal space, and whose radius has an imaginary value.

We still want to form an idea of how the plane is transformed into itself when rotated about an infinitely distant or ideal centre of rotation (§9). In the former case all points describe conic sections that are tangential at infinity. In the latter case they describe conics that touch the fundamental conic at two real points. They include a line in the finite region, the polar of the ideal centre of rotation. This line is displaced along itself, but the other points do not describe parallel lines, as they would in parabolic geometry,

[61] On this point see the discussion given by Battaglini: Sulla geometria imaginaria di Lobatchefsky. *Giornale di Matematiche* 5 (1867).

but conics (becoming flatter in the neighbourhood of the line) touching the fundamental conic at its intersections with the line.

Finally, the *trigonometric formulae* for the present measure are obtained immediately from the following considerations. In §11 we have seen that, on the basis of an imaginary conic in the plane and the choice of constants $c = c_1 i$, $c' = c_1' i = \frac{\sqrt{-1}}{2}$, the trigonometry of the plane has the same formulae as spherical trigonometry when one replaces the sides by sides divided by $2c_1$. The same still holds on the basis of a real conic. Because the validity of the formulae of spherical trigonometry rests on analytic identities that are independent of the nature of the fundamental conic. The only difference from the earlier case is that $c_1 = \frac{c}{i}$ is now imaginary.

The trigonometric formulae that hold for our measure result from the formulae of spherical trigonometry by replacing sides by sides divided by $\frac{c}{i}$.

But this is the same rule one has for the trigonometric formulae of hyperbolic geometry. The constant c is the characteristic constant of hyperbolic geometry. One can say that planimetry, under the assumption of hyperbolic geometry, is the same as geometry on a sphere with the imaginary radius $\frac{c}{i}$.

The preceding immediately gives a model of hyperbolic geometry, in which we take an arbitrary real conic and construct a projective measure on it. Conversely, if the measure given to us is representative of hyperbolic geometry, then the infinitely distant points of the plane form a real conic enclosing us, and the hyperbolic geometry is none other than the projective measure based on this conic.

§13
The special measure in the plane. Parabolic geometry

The measure for parabolic geometry is not among those considered at present, since it does not use a proper conic section as fundamental figure. Moreover, it is a limiting case of the general measure considered so far, where the fundamental conic degenerates to a point pair. This fundamental point pair is imaginary for parabolic geometry; *it consists of the two imaginary points at infinity of the circle.*

An imaginary point pair, as we have mentioned in passing, may be regarded as a bridge between a real and an imaginary conic section, and for this reason parabolic geometry may also be regarded as the transitional case between hyperbolic and elliptic geometry. Suppose, for example, that a hyperbola is given, whose (imaginary) minor axis has a fixed value, while its principal axis gradually shrinks to zero and then becomes imaginary. At the limit zero the two branches of the hyperbola collapse to a doubly-counted line, the minor axis. This line represents the conic section, in so far as it is generated by points. But in so far as it is enveloped by lines, it degenerates to two conjugate imaginary points, which lie at the ends of the constant minor axis on the doubly-counted line. In the limiting process all tangents to the conic section become imaginary tangents to the line which now represents the whole conic, and to which they are double tangents. If the principal axis then becomes imaginary as well, then the conic no longer contains any real elements at all.

However, we shall first consider a measure in the plane that uses a point pair instead of a fundamental conic. Such a measure will be called a *special* measure, in contrast with the *general* measures considered until now. It goes without saying that one could consider the degeneration of a conic into a line pair instead of a point pair; if we confine ourselves to the latter and give it a particular name, it is because it is the case that includes parabolic geometry.

When the fundamental conic section degenerates to a point pair, the determination of angle remains similar to the general case. Each line pencil whose centre is not on the line connecting the two fundamental points, i.e. on the fundamental conic, has two distinct fundamental lines – those going through the fundamental points. On the other hand, the determination of the distance between two points is essentially different from the general case. Since the fundamental conic now consists of a doubly-counted line, it cuts all lines in coincident point pairs. The distance measured by it, as long as the constant c is not infinite, is zero. To make the distance finite we must give c an infinite value. Then the distance becomes an algebraic function of the coordinates. But the previous comparability of line segments and angles is lost; more correctly: segments are only comparable with infinitely small angles. Even when we give c an infinite value, the distance between points whose connecting line goes through a fundamental point is equal to zero. Because these lines correspond to the tangents of the earlier conic section. A zero angle occurs between any two lines which meet on the line connecting the two fundamental points.

We give the name circles to the conic sections passing through the two fundamental points; these include concentric circles, which touch at the two fundamental points. Each system of concentric circles includes one with radius ∞. It degenerates into the doubly-counted line connecting the two fundamental points. *Thus the points at infinity now form a doubly-counted line.* Circles no longer have a self-dual meaning. Those lines that cut a given line at a constant angle no longer envelop a proper circle, but an infinitely distant point. The circle with centre at infinity, which has four-point contact with the fundamental conic at the centre, now breaks into the line at infinity and another line, etc. All of these are things which follow immediately from the previous situation by passing to the limit.

Just as we were able to give the name *motions* to a triply-infinite family of linear transformations fixing a conic section, so can we here. However, it no longer suffices to define motions as those linear transformations (or rather, a class of them) fixing the fundamental figure. Because a point pair is fixed not just by a triply-infinite family of linear transformations of the plane, but by a quadruply-infinite family. Among them, however, a triple infinity is distinguished by the property that each of them preserves the circles of a concentric pencil. This family subdivides further into two triply-infinite families. One consists of the motions, the other consists of transformations that carry plane figures to congruent inverses. The difference between the two families is that the motions fix both fundamental points, while the other transformations exchange them. Each motion of the plane is a rotation about a point. If the motion is a translation, i.e. if the centre of rotation is at infinity, then all points of the plane describe parallel lines – lines which meet at the same point at infinity. There now exists the concept of *direction; parallel lines have the same direction.* The motions have lost the self-dual character they had in the general case. Besides the relationship of *congruence*, defined by the triple infinity of motions, and that of *inverse congruence*, defined by the triple infinity of transformations in the second family, there is also *direct and inverse similarity*, defined by the quadruple infinity of linear transformations preserving the fundamental figure. A similarity is direct if the fundamental points are fixed, and inverse if they are exchanged. Under a similarity all angles are unaltered, while distances are multiplied by a constant. It should be noted that now we can reach all points of the plane by motions, except the points on the line at infinity. There is no ideal region, as in the case of a real conic section, or, if one prefers, it has collapsed to the doubly-counted boundary.

The analytic formula now representing the distance between two points – and we shall confine ourselves to this – takes the following form. Let $p_x = p_1x_1 + p_2x_2 + p_3x_3 = 0$ be the equation of the line at infinity. Also let $P_{xy} = 0$ be the condition under which the line connecting (x) and (y) goes through one of the fundamental points. Then the distance between two points becomes

$$\frac{c\sqrt{P_{xy}}}{p_x p_y}.$$

Thus the distance between two points is an algebraic function of their coordinates.

In fact one derives this formula by taking a limit of the general expression for distance. The general expression may be written

$$2ic \arcsin \frac{\sqrt{\Omega_{xy}^2 - \Omega_{xx}\Omega_{yy}}}{\sqrt{\Omega_{xx}\Omega_{yy}}}.$$

If now $\Omega = 0$ degenerates to a point pair, then $\Omega_{xy}^2 - \Omega_{xx}\Omega_{yy}$ becomes identically zero, but in such a way that it contains a vanishing constant factor (the discriminant of Ω). If the latter is cancelled, what remains of $\Omega_{xy}^2 - \Omega_{xx}\Omega_{yy}$ is precisely P_{xy}, i.e. the expression whose vanishing is equivalent to the tangency of the line through (x) and (y) with the degenerate conic. But because of the vanishing factor we can replace the *arcsine* by the *sine* itself, and by combining the vanishing factor with $2ic$ to form a new constant C, and finally writing p_x^2 and p_y^2 in place of Ω_{xx} and Ω_{yy} (since $p^2 = 0$ is the equation of the degenerate conic in point coordinates), we arrive at the given expression.

From it we obtain the usual expression for the distance between two points in parabolic geometry by using the usual representation of the two fundamental points. In this notation the line at infinity has equation: constant $= 0$. Thus $p_x = p_y$ is a constant k. The circular points on it are represented in rectangular coordinates as its intersection with the line pair

$$x^2 + y^2 = 0.$$

The condition for two points (x, y) and (x', y') to lie so that their connecting line goes through a circular point then is:

$$(x - x')^2 + (y - y')^2 = 0.$$

Consequently the distance between the two points is

$$\frac{C}{k^2} \sqrt{(x - x')^2 + (y - y')^2}.$$

Finally, if x and y are replaced by those multiples that make distance on the x-axis or y-axis exactly equal to the difference between coordinates, we get

$$\sqrt{(x - x')^2 + (y - y')^2},$$

the usual expression for distance in rectangular coordinates.

We shall not discuss here how the ideas of parabolic geometry with its imaginary fundamental points fit into the previous general considerations.[62] We only want to stress that with imaginary fundamental points the trigonometric formulae become the relevant formulae of parabolic geometry, so the angle sum of a triangle is exactly π, whereas with a real fundamental conic it is smaller, and with an imaginary fundamental conic it is larger.

§14
Special measures in the plane tangential to a general measure. Curvature of the latter

Just as we were able, in §7, to give a special measure on the line, coinciding with a general measure in the neighbourhood of a point, and tangential to

[62] Cf. Cayley, l.c.

it, so we can also speak of a special measure on the plane which is tangential to a given general measure at a point. The latter (§7) has as line at infinity the polar of the given point with respect to the fundamental conic of the general measure, and as fundamental points the two contact points of the tangents from it to the fundamental conic. Then, with a suitable choice of constants, both measures yield the same angle measure at the given point, and they agree up to first order on the distances between the given point and those infinitely close to it. Circles with centre at the given point of contact in the general measure, i.e. conic sections which touch the fundamental conic at the two fundamental points of the tangential special measure, are also circles with respect to the latter. In particular, the fundamental conic itself, which is the circle of infinite radius for the general measure, is also a circle for the special measure, but a circle of finite radius. One finds the magnitude of this radius to be the constant $2c$. This is because the given general and tangential special measures determine two like measures on each of the lines through the point of contact which are also tangential. And the fundamental points of the general measure on a line are the intersections of the line with the fundamental conic, whose distance apart, according to the tangential special measure (§7) is equal to $4c$. For this reason, the required radius is $2c$.

We now want to focus particularly on the two cases of the general measure considered in §11 and §12 as models of elliptic and hyperbolic geometry, namely those where the fundamental conic is imaginary and where it is real and enclosing us.

In both cases the tangential special measure has imaginary fundamental points, since the polar of the point of contact does not meet the fundamental conic at real points. But there is a difference between the two kinds of general measure, analogous to that found in §7 for the corresponding measures on the line. If the fundamental conic is imaginary, then the special measure runs ahead, i.e. the distance of a point from the contact point is always greater in the special measure than the general measure. Conversely, if the fundamental conic is real, the special measure stays behind the general.[63] This running ahead, or staying behind, of the special measure will be called *curvature* of the general measure, in the former case *positive*, and in the latter case *negative*. As a *measure of curvature* we take the same expression as in §7 giving the curvature of the general measure on a line through the point of contact, namely $-\frac{1}{4c^2}$. This expression is independent of the contact point originally chosen, and of the line taken through it. Thus we have the theorem:

The curvature of the general measure is the same at all points, and equal

[63] Of course this holds only for points inside the fundamental conic; for the points outside the measure can run ahead or stay behind depending on the direction in which one moves.

to $-\frac{1}{4c^2}$.

It is positive for an imaginary fundamental conic (and hence for elliptic geometry), and negative for a real fundamental conic (and hence for hyperbolic geometry).

For the transitional case where the fundamental conic degenerates to an imaginary point pair (parabolic geometry), the curvature is zero.

It will now be shown that the present definition of curvature of a plane measure is equivalent to the one given by Gauss for the curvature of two-dimensional manifolds. The only difference between the concept of curvature here and in Gauss is that for Gauss the curvature is a permanent property of the geometric figure in question, whereas here it is only a property of the measure that happens to be chosen for the given figure, the plane.

As is well known, the Gaussian curvature is computed from the expression for the square of the element of arc:

$$ds^2 = Edu^2 + 2Fdudv + Gdv^2.$$

First we shall set up this expression for the present case. Let $\Omega = 0$, as always, be the fundamental conic section. Let Ω_{xx} have its previous meaning, and let $\Omega_{x,dx}$, $\Omega_{dx.dx}$ denote the expressions that result from Ω_{xy} and Ω_{yy} by introduction of the differential dx in place of y. Now the distance between two points (x) and (y) was

$$2ic \arcsin \frac{\sqrt{\Omega_{xx}\Omega_{yy} - \Omega_{xy}^2}}{\sqrt{\Omega_{xx}\Omega_{yy}}}.$$

If one sets $y_a = x_a + dx_a$ then this expression becomes

$$2ic \arcsin \frac{\sqrt{\Omega_{xx}\Omega_{dx,dx} - \Omega_{x,dx}^2}}{\Omega_{xx}}$$

after omission of higher order terms. Or, if we replace the *arcsin* for small arguments by the *sine* itself,

$$2ic \frac{\sqrt{\Omega_{xx}\Omega_{dx,dx} - \Omega_{x,dx}^2}}{\Omega_{xx}}.$$

Thus the square of the element of *arc* is

$$ds^2 = 4c^2 \frac{\Omega_{x,dx}^2 - \Omega_{xx}\Omega_{dx,dx}}{\Omega_{xx}^2}.$$

We want to bring this expression into a simpler form by special choice of coordinates. Since the fundamental conic is a circle for the tangential special measure, and since also the fundamental points of the latter are imaginary – as for the usual parabolic measure – we can write the equation of the fundamental conic in the ordinary form for a circle:

$$x^2 + y^2 = 4c^2.$$

This equation concerns the coordinates x, y for the tangential special measure, because the radius of the fundamental circle, according to this measure, is $2c$.

Now

$$\Omega_{xx} = x^2 + y^2 - 4c^2, \quad \Omega_{dx,dx} = dx^2 + dy^2, \quad \Omega_{x,dx} = xdx + ydy.$$

Hence the expression for the square of the element of *arc* is:

$$\begin{aligned} ds^2 &= 4c^2 \frac{(xdx + ydy)^2 - (x^2 + y^2 - 4c^2)(dx^2 + dy^2)}{(x^2 + y^2 - 4c^2)^2} \\ &= 4c^2 \frac{-(ydx - xdy)^2 + 4c^2(dx^2 + dy^2)}{(x^2 + y^2 - 4c^2)^2} \end{aligned}$$

If one now introduces new variables (polar coordinates for the special measure) by setting:

$$x = r\cos\varphi, \quad y = r\sin\varphi,$$

the square of the element of *arc* becomes

$$ds^2 = \frac{16c^4 dr^2}{(r^2 - 4c^2)^2} - \frac{4c^2 r^2 d\varphi^2}{r^2 - 4c^2},$$

an expression which becomes the ordinary expression for the element of arc in polar coordinates as c goes to infinity.[64] If one compares it with the basic

[64] If one takes r to be constant, then

$$ds = \frac{2cr}{\sqrt{4c^2 - r^2}} d\varphi.$$

Thus the circumference of a circle with radius r equals $\frac{4cr\pi}{\sqrt{4c^2 - r^2}}$. But this r is only the radius of the circle measured in the special measure tangential to the centre. The radius ρ in the general measure is obtained from the formula in the text by writing $d\rho$ in place of ds and setting $d\varphi$ equal to zero, so

$$d\rho = \frac{-4c^2 dr}{r^2 - 4c^2}$$

or

$$r = 2c\frac{e^{\frac{\rho}{c}} - 1}{e^{\frac{\rho}{c}} + 1}.$$

If one substitutes this for r, then one finds the circumference of the circle with radius ρ to be

$$2c\pi(e^{\frac{\rho}{2c}} - e^{-\frac{\rho}{2c}}),$$

a formula which Gauss gave in a letter to Schumacher. The constant k that he used there corresponds precisely to the constant c used here.

formula of Gauss:

$$ds^2 = Edu^2 + 2Fdudv + Gdv^2,$$

then F vanishes, and E and G depend on only one variable, say u. But under this hypothesis the Gaussian curvature K is given by

$$4E^2G^2K = E\left(\frac{\partial G}{\partial u}\right)^2 + G\frac{\partial E}{\partial u}\frac{\partial G}{\partial u} - 2EG\frac{\partial^2 G}{\partial u^2}.$$

If one gives E, G their values

$$E = \frac{16c^4}{(u^2-4c^2)^2}, \quad G = -\frac{4c^2u^2}{u^2-4c^2},$$

then one gets

$$K = -\frac{1}{4c^2}$$

– *the same value that we found previously.*

We can now state the theorem in terms of curvature in the Gaussian sense:

According as we impose elliptic, hyperbolic or parabolic geometry, the plane becomes a surface of constant positive, constant negative, or zero curvature.

It is also for this reason (as mentioned in §1) that when the parabolic measure is used a basis, elliptic geometry finds its interpretation on the sphere or deformations of it, and hyperbolic geometry finds its interpretation on the surfaces of constant negative curvature.

§15
The relationship between elliptic, hyperbolic and parabolic geometry in the plane

In the preceding we have seen how both the measure for parabolic geometry, and those for elliptic and hyperbolic geometry in the plane, are special cases of the general projective plane measure. Parabolic geometry uses the fundamental conic consisting of a pair of imaginary points, the so-called imaginary circular points at infinity.[65] The locus of points at infinity is a doubly-counted line. Elliptic geometry uses a fundamental conic which is proper, but imaginary. Finally, hyperbolic geometry has a fundamental conic which is both proper and real (and which encloses us).

[65] To say these points are at infinity is not strictly justified, since their distance from a point at finite distance is not infinite, but indeterminate, since indeed all circles around the finite point contain the two circular points at infinity.

In the neighbourhood of a point, as we have just seen, all three geometries have coincident measures, whether parabolic, elliptic or hyperbolic. They are tangential at the point in question, and parabolic geometry gives a tangential special measure for both elliptic and hyperbolic geometry.

Thus if we are actually given parabolic geometry we can immediately construct a geometry which models hyperbolic geometry by constructing a general measure with real fundamental conic, tangential to the given special measure at a point of our choice. We achieve this by describing a circle of radius $2c$ centred on our point, and using it as the basis for a projective measure with the constant c determining the distance between two points and the constant $c' = \frac{\sqrt{-1}}{2}$ determining the angle between two lines. This general measure approaches the given parabolic measure more closely as c becomes larger, coinciding with it completely when c becomes infinite.

In a similar way we construct a geometry that shows how elliptic geometry can tend toward the parabolic. To do this it suffices to give a pure imaginary value $c_1 i$ to the c we used previously. Then we fix a point at distance $2c_1$ above the given point of contact and take the distance between two points of the plane to be c_1 times the angle the two points subtend at the fixed point. The angle between two lines in the plane is just the angle they subtend at the fixed point. The resulting measure approaches more closely to the parabolic measure the greater c_1 is, and becomes equal to it when c_1 is infinite.

When elliptic or hyperbolic geometry is actually the given geometry one can in this way make a model presenting its relationship with parabolic or the other geometry.

It now remains only to carry over these ideas to figures of the first or second kind in space, which will be done as briefly as possible.

§16
The projective measure in space

The general projective measure in space is based on a *fundamental surface of second degree.*

To determine the distance between two points, one connects them by a straight line. The latter meets the fundamental surface in two new points, which have a certain cross-ratio with the two given points. *The logarithm of this cross-ratio, multiplied by an arbitrary constant c, is called the distance between the two given points.*

In a similar way one determines the angle between two given planes. Through their line of intersection, one takes the two tangent planes to the fundamental surface. The latter determine a certain cross-ratio with the two given planes. *The angle between the two planes is equal to the logarithm of this cross-ratio, multiplied by an arbitrarily chosen constant c'.*

A *motion* of space is understood to be a linear transformation that preserves the fundamental surface. A surface of second degree is preserved by a sextuple infinity of linear transformations. But these fall into two classes, one of which is a closed system, and the other not [without now having to specify whether the points of the surface are real]. The two classes may be characterised by their effect on the generators of the surfaces. Transformations in the first class – which we call the motions[66] of the space – leave the system of rectilinear generators fixed; transformations in the second class permute them. There is a sextuple infinity of motions, and they leave the metric relations unaltered.

By *spheres* one means those surfaces of second degree that touch the fundamental surface along a plane curve. The centre of the sphere is the pole of the plane containing the curve of contact. The fundamental surface itself is a sphere of radius ∞ about an arbitrary centre, etc.

By attending to the real elements of the space one can decide whether the fundamental surface is imaginary or real, and in the latter case whether it is ruled or not.

If the fundamental surface is *imaginary*, then all lines have a finite length, and all plane pencils of lines have a finite angle sum. This case includes the measure for *elliptic* geometry when the constant c' for angle determination is $\frac{\sqrt{-1}}{2}$, so that the angle sum of a plane pencil is π.

The case where the fundamental surface is *real* and *ruled*, so that it is a one-sheeted hyperboloid, we shall not consider further, because it is not related to the three geometries considered here – elliptic, hyperbolic and parabolic.

Finally, if the fundamental surface is *real* and *not ruled* then the points inside it acquire a measure which becomes the measure for *hyperbolic* geometry when one again sets the constant c' equal to $\frac{\sqrt{-1}}{2}$.

Parabolic geometry occurs as a special case of the general measure that enters when the fundamental surface is particularised to a conic section, in particular an imaginary conic section. The fundamental conic of parabolic geometry is the so–called imaginary circle at infinity. The undualistic properties of the parabolic measure have their origin in the undualistic character of the particularisation experienced by the fundamental surface.

Again one can speak of the *curvature* of a general measure, etc.; however, all these things will be omitted here for the sake of brevity.

[66]I have already explained this behaviour in an earlier work: Über die Mechanik starrer Körper, *Math. Annalen* 4. I must add that Schering has already considered the motions of space, in the sense of hyperbolic geometry, in the essay: Die Schwerkraft in Gausschen Raume, *Gött. Nachrichten* 1870, no. 15.

§17
The independence of projective geometry from the theory of parallels

Against all of the preceding there is a possible objection, not unfounded at this stage, but easily removed.

The construction of the general projective measure was based on a geometric process, in which we defined the distance between two points, etc. as logarithms of certain cross-ratios, which were based in turn on homogeneous coordinates. Both things – the cross-ratio and the homogeneous coordinates – depend on the parabolic measure in their usual formulation, where the cross-ratio, as well as the homogeneous coordinates are defined as the ratios of certain segments. Thus if the given measure is not in fact parabolic one cannot speak of these things, and all the preceding discussion loses its validity.

On the other hand, one can convince oneself that projective geometry is valid, independently of the nature of the measure.

The proof of this can be carried out by constructing projective geometry once on the basis of elliptic geometry, and again on the basis of hyperbolic geometry. This is not hard to do, as one can see that projective geometry holds for their points, line- and plane pencils in space as well as it does in parabolic geometry with an elliptic measure.

However, it is more important to notice that *projective geometry can be developed without the construction of a measure.*

In order to prove the validity of projective geometry in any space, it suffices to make constructions in this space which involve only the so-called positional [incidence] relations and which do not go outside the space. Naturally the cross-ratio cannot be defined as a ratio of lengths, since this assumes a measure to be known. But von Staudt's *Beiträge zur Geometrie der Lage*[67] gives the necessary materials to define the cross-ratio as a pure number. With the cross-ratio established, we can then set up homogeneous point and plane coordinates, since these are none other than the relative values of certain cross-ratios, as von Staudt[68] and more recently Fiedler[69] have shown. It remains open whether all real values of coordinates are realised by actual elements of the space. If this is not the case, then nothing stands in the way of adjoining, to the proper elements, ideal elements corresponding to the coordinate values in question. This happens in parabolic geometry when we speak of the plane at infinity. To construct hyperbolic geometry

[67] §27, Nr. 393.

[68] *Beiträge* §29, Nr. 411.

[69] Vierteljahrsschrift der naturforschenden Gesellschaft in Zürich. XV.2 (1871). Fiedler's *Darstellende Geometrie.* Leipzig 1871.

one has to adjoin a whole piece of space. However, in elliptic geometry no improper elements need to be adjoined.

§18
Derivation of the three geometries – elliptic, hyperbolic and parabolic – from the projective

If one has set up projective geometry, as explained above, then one can construct the general Cayley measure. The measure is invariant under a sixfold infinity of linear transformations that we call motions of the space, and can be regarded as generated by them (§§2, 3).

Now consider the actual motions in space and the measure based on them. One sees, conversely, that the sixfold infinity of motions are linear transformations. They also preserve a surface, the surface of points at infinity. However, it is easy to prove that no surfaces are preserved by a sixfold infinity of linear transformations except surfaces of second degree and their degenerations. Thus the points at infinity form a surface of second degree, and the motions of the space are included in the above sixfold infinity of linear transformations which preserve a surface of second degree. For this reason, the measure given by motions is subsumed under the general projective measure. While the latter relates to an arbitrary surface of second degree, the former has this surface given once and for all.

The nature of this second degree surface underlying the actual measure can be determined more precisely. Notice that continued rotation of a plane about one of its axes in the finite region eventually brings it back to its initial position. This means that the two tangent planes from a line in the finite region to the fundamental surface are imaginary. If they were real, the pencil of planes through the line would include two real planes at infinity (i.e., planes that make an infinite angle with all the others) and then no continued rotation of a plane in the same sense could bring it back to its initial position.

Given that these two planes are imaginary, or equivalently, that the tangent sphere to the fundamental surface from an arbitrary point (accessible to us by motion) is imaginary, exactly three cases are conceivable:

1. *The fundamental surface is imaginary.* This gives elliptic geometry.

2. *The fundamental surface is real, not ruled and it encloses us.* The assumption of hyperbolic geometry.

3. (Transitional case.) *The fundamental surface degenerates to an imaginary plane curve.* The hypothesis of ordinary parabolic geometry.

Thus we are led to just the three geometries that have been set up from entirely different considerations, as reported in §1.

Düsseldorf, 19 August 1871.

Excerpt from Klein's second paper of the same name (Math. Ann. 6, 1873)

The investigation of noneuclidean geometry is by no means intended to decide the validity of the parallel axiom, but only *whether the parallel axiom is a mathematical consequence of the remaining axioms of Euclid*; a question to which these investigations give a definite *no*. Because they have shown that these remaining axioms suffice to construct a system of theories which includes euclidean geometry merely as a special case.

Translator's Introduction

Poincaré's *Theory of fuchsian groups,* *Memoir on kleinian groups,* *On the applications of noneuclidean geometry to the theory of quadratic forms.*

Hyperbolic geometry took a new turn with the advent of Poincaré. Beltrami and Klein were primarily geometers, who used known geometry to construct realisations of what was, until then, an unknown geometry. Poincaré began in other fields, and made the surprising discovery that hyperbolic geometry was *already present* in mainstream mathematics. In a famous passage, he described how the revelation came to him:

> Just at this time I left Caen, where I was then living, to go on a geological excursion under the auspices of the school of mines. The changes of travel made me forget about my mathematical work. Having reached Coutances, we entered an omnibus to go some place or other. At the moment when I put my foot on the step the idea came to me, without anything in my former thoughts seeming to have paved the way for it, that the transformations I had used to define Fuchsian functions were identical with those of non-Euclidean geometry. I did not verify the idea: I should not have had time, as, upon taking my seat in the omnibus, I went on with a conversation already commenced, but I felt a perfect certainty. On my return to Caen, for conscience' sake I verified the result at my leisure.
>
> Then I turned my attention to the study of some arithmetic questions apparently without success and without a suspicion of any connection with my previous researches. Disgusted with my failure, I went to spend a few days at the seaside, and thought of something else. One morning, walking along the bluff, the idea came to me, with just the same characteristics of brevity, suddenness and immediate certainty, that the arithmetic tranformations of ternary quadratic forms were identical with those of non-Euclidean geometry.
>
> Poincaré [1960]

This discovery cast hyperbolic geometry in an entirely new light. Now it could be seen as not merely logically valid, but also natural, familiar and potentially useful. Poincaré proceeded to show how useful it was with a slew of applications to complex analysis, differential equations, number theory and later topology.

To make these applications, he developed new models of hyperbolic geometry, which are the subject of the three translations below. The first two translations are excerpts from long papers Poincaré wrote on the group theory of differential equations. The third, which is not so well known, is a complete paper on applications of hyperbolic geometry to number theory. As we shall see, number theory contains possibly the earliest hints of hyperbolic geometry in mainstream mathematics, some being visible in the memoir of Lagrange [1773] on quadratic forms.

Theory of fuchsian groups

Poincaré first met hyperbolic geometry when attempting to understand a new kind of "periodicity" occurring in solutions of certain differential equations. Classical periodic functions, like cos and sin, have a *period* 2π such that

$$\cos(x + 2\pi) = \cos x,$$
$$\sin(x + 2\pi) = \sin x.$$

In other words, they are invariant under the substitution of $x + 2\pi$ for x, and hence invariant under the whole *group* of substitutions $x \mapsto x + 2n\pi$ for $n \in \mathbb{Z}$.

Almost as well known, in Poincaré's time, were the doubly-periodic, or *elliptic* functions. An elliptic function is invariant under the group of substitutions $z \mapsto z + m\omega_1 + n\omega_2$ for $m, n \in \mathbb{Z}$, where ω_1 and ω_2 are nonzero complex numbers with different arguments. Double periodicity is easily understood with the help of a tessellation of the euclidean plane $\mathbb{C}$ by parallelograms whose vertices are the points $m\omega_1 + n\omega_2$. The elliptic function takes the same value at corresponding points in all the parallelograms, for example, at all the points marked * in Figure 4.1.

The new kind of "periodic" function, which Poincaré called *fuchsian* because a large class of them were discovered by Fuchs, is invariant under a group of substitutions of the form

$$z \mapsto \frac{az + b}{cz + d}, \quad \text{where} \quad ad - bc \neq 0.$$

Fuchsian functions and their groups originated from certain differential equations, the full story of which may be found in Gray [1986]. Poincaré wished

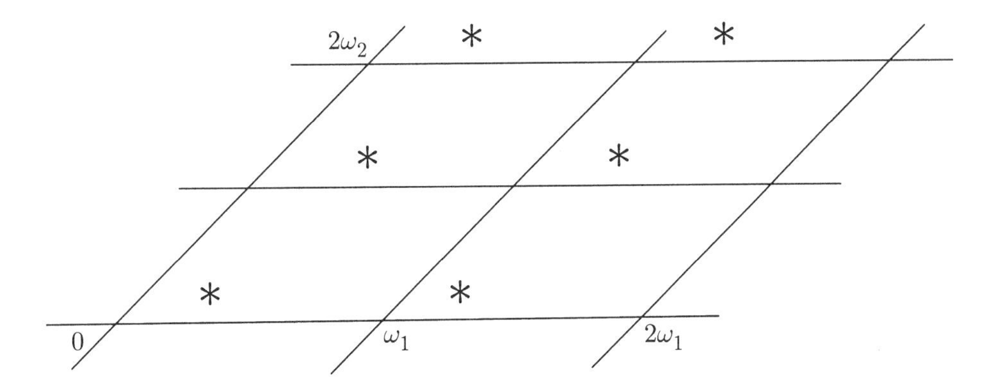

Figure 4.1: Double periodicity

to study these *fuchsian groups* by means of tessellations, and indeed there was already a published precedent in the work of Schwarz [1872], to which Poincaré refers in the historical remarks at the end of his paper. (These remarks are not included here, but see Poincaré [1985] for a full translation.) Schwarz gave the beautiful tessellation picture shown in Figure 4.2.

The tessellation consists of the curvilinear triangles inside the circle, which make up right-angled pentagons. The pentagons are obviously not all the same size, however the tessellation has at least abstract symmetry, because it is mapped onto itself by a group of functions of the form

$$z \mapsto \frac{az+b}{cz+d}.$$

Moreover, such *linear fractional* functions do preserve certain geometric properties: in particular, angles and circles.

It was around this point in the mathematics that Poincaré had his wonderful insight: linear fractional functions can be used to *define* a new concept of length, under which the cells of the corresponding tessellation are of equal size. *The corresponding geometry is Bolyai-Lobachevsky geometry* (which we now call *hyperbolic*).

Once guessed, this fact is easily verified, particularly if the disc is first mapped onto the upper half plane by a function such as $z \mapsto (1-zi)/(z-i)$. The linear fractional functions that map the upper half plane onto itself are of the form

$$z \mapsto \frac{az+b}{cz+d}, \quad \text{where} \quad a,b,c,d \in \mathbb{R} \quad \text{and} \quad ad-bc>0.$$

The condition $ad-bc \neq 0$ guarantees invertibility, and $ad-bc>0$ that the upper half plane is mapped onto itself (rather than onto the lower half

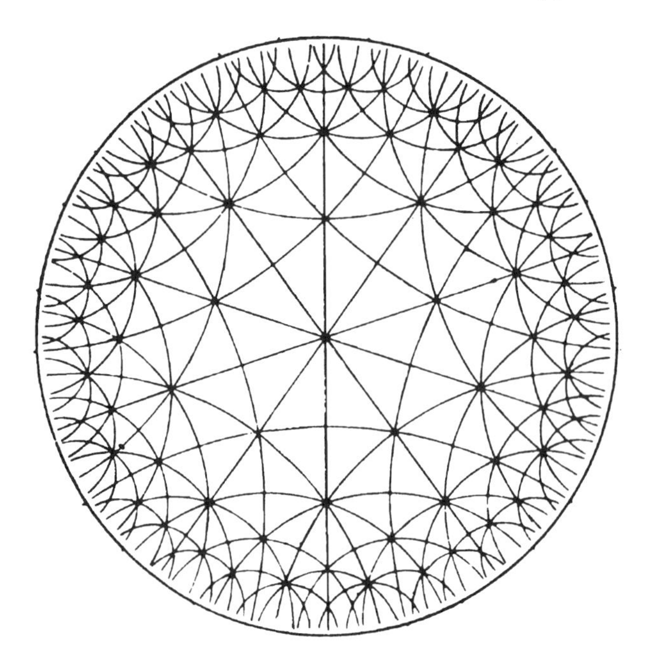

Figure 4.2: Schwarz's tessellation

plane). Such a function is composed from functions of the form

$$z \mapsto rz \quad \text{for} \quad r > 0,$$
$$z \mapsto z + s \quad \text{for} \quad s \in \mathbb{R},$$
$$z \mapsto -1/z,$$

each of which is easily seen to preserve the line element

$$ds = \frac{\sqrt{dx^2 + dy^2}}{y} = \frac{|dz|}{y}$$

used by Beltrami for his half plane model of hyperbolic geometry (see the second paper in this volume, equation (21)).

This is why Poincaré begins with real linear fractional transformations of the upper half plane, and defines length as the integral of $|dz|/y$. It follows that each real transformation $z \mapsto (az + b)/(cz + d)$ is length-preserving, or an *isometry* as we now say. His other claims, such as the fact that hyperbolic "lines" are semicircles with centres on the real axis, are also easy to prove, and are now part of the standard approach to hyperbolic geometry. (See for example Beardon [1983].)

Memoir on kleinian groups

The interpretation of invertible real linear transformations as isometries of the hyperbolic plane does not extend to all linear fractional transformations;

there are simply too many of them. For any two point pairs $z_1 \neq z_2$, $z_3 \neq z_4$ there is an invertible linear fractional transformation mapping z_1, z_2 onto z_3, z_4 respectively – and we cannot allow any two points to be the same distance apart. It seems that a geometric interpretation of all the transformations $z \mapsto (az+b)/(cz+d)$ is doomed, but Poincaré escaped this conclusion with a splendid piece of lateral thinking: he moved *off the plane*, into space.

Just as the real axis can be viewed as the "boundary at infinity" of the half plane with line element $\sqrt{dx^2 + dy^2}/y$, the (x, y)-plane can be viewed as the boundary at infinity of the half (x, y, w)-space with line element $\sqrt{dx^2 + dy^2 + dw^2}/w$. (I use w for the third coordinate to avoid a clash with $z = x + iy$.) This is Beltrami's half space model of *hyperbolic space* (second paper, equation (21) again). It now turns out that each invertible transformation $z \mapsto (az + b)/(cz + d)$ can be viewed as the "shadow," on this boundary at infinity, of a motion of hyperbolic space.

The idea is to compose each invertible function $z \mapsto (az + b)/(cz + d)$ from the simple functions

$$
\begin{aligned}
&z \mapsto rz, \quad z \mapsto e^{i\theta} z \quad r, \theta \in \mathbb{R} \\
&z \mapsto z + s, \\
&z \mapsto 1/z,
\end{aligned}
$$

and then to express each of the latter as a product of *inversions* ("reflections in circles"). Inversion in a circle C sends each $z \in \mathbb{C}$ to the $z' \in \mathbb{C}$ on the same ray through the centre c of C, but at distance such that

$$|z - c||z' - c| = (\text{radius of } C)^2.$$

Ordinary reflection in a line is the limiting case of inversion as the radius tends to ∞. Another important case is inversion in the unit circle, which is $z \mapsto 1/\bar{z}$. It is easily checked that each of the simple functions is the product of two inversions, hence every invertible function $z \mapsto (az + b)/(cz + d)$ is the product of an even number of inversions.

Poincaré extends these inversions in the obvious way to inversions in *spheres*, orthogonal to the (x, y)-plane and passing through the circles of inversion. This extends the map $z \mapsto (az+b)/(cz+d)$ to a map of hyperbolic space, and the latter map is easily seen to be an isometry (off the (x, y)-plane, of course) since inversion in a sphere preserves hyperbolic distance. In fact, since the spheres are orthogonal to the boundary of the half-space, they model hyperbolic planes in space, and inversion is actually hyperbolic reflection in a plane.

Thus groups of general linear fractional transformations can be viewed as groups of isometries of hyperbolic space. The type that Poincaré was interested in, and which he called *kleinian groups*, are symmetry groups of tessellations of hyperbolic space. Hence he was able to study them the same way he studied fuchsian groups, by studying tessellations.

On the applications of noneuclidean geometry to the theory of quadratic forms

The theory of quadratic forms has ancient roots (for example, in Pythagoras' theorem), but its depth and importance were first recognised by Fermat. He made a series of claims about primes of the form x^2+y^2, x^2+2y^2 and x^2+3y^2 which sparked the interest of Euler and Lagrange, and led to the theory of *binary* quadratic forms $ax^2 + bxy + cy^2$, where a, b and c are integers. A typical result of the theory is Fermat's [1640] claim that the primes of the form $x^2 + y^2$ are precisely those of the form $4n + 1$.

Lagrange [1773] developed a systematic appproach to such questions by considering the transformations of a form $ax^2 + bxy + cy^2$ induced by change of variables:

$$x' = qx + ry,$$
$$y' = sx + ty.$$

The resulting form $a'x'^2 + bx'y' + cy'^2$ represents exactly the same integers as $ax^2 + bxy + cy^2$ if the pairs (x', y') include all pairs of integers. It is easy to show that this is the case if and only if

$$\begin{vmatrix} q & r \\ s & t \end{vmatrix} = qt - rs = \pm 1.$$

Forms related by such a transformation are said to be *equivalent.*

The link with hyperbolic geometry grows out of this concept of equivalence, and especially out of corresponding *group* of transformations, the group of integer 2×2 matrices with determinant ± 1. The same group occurs as the group of linear mappings of any *plane lattice* $\langle \omega_1, \omega_2 \rangle$ onto itself, where we define

$$\langle \omega_1, \omega_2 \rangle = \{m\omega_1 + n\omega_2 : m, n \in \mathbb{Z}\},$$

for any ω_1, ω_2 with different arguments. Then if

$$\omega_1' = q\omega_1 + r\omega_2,$$
$$\omega_2' = s\omega_1 + t\omega_2,$$

we find that

$$\langle \omega_1', \omega_2' \rangle = \langle \omega_1, \omega_2 \rangle \Leftrightarrow q, r, s, t \in \mathbb{Z} \text{ and } \begin{vmatrix} q & r \\ s & t \end{vmatrix} = \pm 1.$$

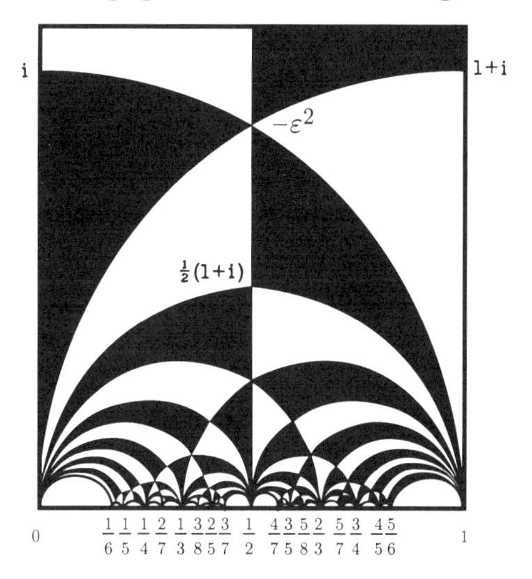

Figure 4.3: The modular tesselation (from P. du Val, *Elliptic Functions and Elliptic Curves*, ©1973, Cambridge University Press. Reprinted with the permission of Cambridge University Press).

This property is obviously relevant to the theory of elliptic functions with periods ω_1, ω_2, and in fact it is due to this connection that the group of 2×2 integer matrices with determinant ± 1 is called the (extended) *modular* group. The word comes from the *modulus* $\omega_1/\omega_2 = \tau$, a complex number which captures the shape of the parallelogram with sides ω_1 and ω_2.

The standard picture of the modular group comes from viewing the modulus τ as a point of the upper half plane, and the transformation $(\omega_1, \omega_2) \mapsto (\omega_1', \omega_2')$ as the map $\tau \mapsto \tau'$. Now if

$$\omega_1' = q\omega_1 + r\omega_2,$$
$$\omega_2' = s\omega_1 + t\omega_2,$$

it is easily checked that

$$\tau' = \frac{q\tau + r}{s\tau + t}.$$

Thus we are viewing the modular group as a group of hyperbolic isometries. Also, the group acts *discontinuously* on the half plane: each τ in the half plane has a neighbourhood containing no equivalent point τ'. This leads to a tiling of the half plane by regions of inequivalent points, and the modular group is the symmetry group of the tiling – the famous modular tessellation shown in Figure 4.3.

This tessellation, or at least the region lying above the unit circle, has a number-theoretic meaning probably understood by Gauss (see Scharlau and Opolka [1984], p. 92). When quadratic forms are suitably interpreted as complex numbers, those lying in this region correspond to forms that Lagrange called *reduced.* In terms of parallelogram shapes, $\tau = \omega_1/\omega_2$ is such that $|\tau| \geq 1$ and $\pi/3 \leq 2\pi/3$ so the parallelogram is one for which $|\omega_1| \geq |\omega_2|$ and the angle between side ω_1 and side ω_2 is $\geq \pi/3$ and $< 2\pi/3$.

Figure 4.4: The hyperboloid model and the conformal disc (reproduced here with permission of the artist, Konrad Polthier).

Similar results apply to ternary quadratic forms, such as $x^2 + y^2 - z^2$, when the transformations are restricted to those that preserve the form, or "similarity substitutions" as Poincaré calls them. (In the case of the form $x^2 + y^2 - z^2$, this means transformations which convert it to $x'^2 + y'^2 - z'^2$.) It is still possible, as Poincaré shows, to interpret the transformations as a discontinuous group of isometries of the hyperbolic plane. The group can then be studied via the polygons in the tessellation, one of which represents the class of "reduced" forms.

The special aspect of ternary forms is that they led Poincaré to a new model of the hyperbolic plane, now called the *hyperboloid model*. A thorough treatment of this model may be found in Reynolds [1993], or in the textbook of Ryan [1986]. See Reynolds in particular for information on earlier versions of the model by Killing and Weierstrass. The beautiful illustration of the hyperboloid model shown in Figure 4.4 is due to Konrad Polthier of the Technical University of Berlin. It also shows the conformal disc model, which results from projection of the hyperboloid from the apex of the lower sheet, analogous to stereographic projection of the sphere.

The hyperboloid is only implicit in the paper below, but Poincaré described it explicitly in a later comment:

> One of the most important problems in the subject of indefinite ternary quadratic forms is the study of the discontinuous groups consisting of similarity substitutions, that is, substitutions that preserve the form. Let $F(x, y, z)$ be an indefinite quadratic form.
>
> We can choose the constant K so that $F(x, y, z) = K$ represents a hyperboloid of two sheets. The similarity substitutions then map a point on the hyperboloid to another point on the same sheet and, since the group is discontinuous, the hyperboloid becomes partitioned into infinitely many curvilinear polygons whose sides are diametric sections of the surface. [Intersections of the surface with planes through the origin.] A similarity substitution changes each polygon into another. We now take a perspective view by placing the eye at an umbilic of the surface and taking the plane of projection to be a circular section. One sheet of the hyperboloid is projected inside a circle, and the polygons drawn on this sheet project to polygons bounded by circular arcs of the kind we have discussed in the theory of fuchsian groups. Thus the study of similarity substitutions of quadratic forms reduces to that of fuchsian groups, which is an unexpected rapprochement between two very different theories, and a new application of non-euclidean geometry.
>
> Poincaré, *Œuvres*, vol. 5, p. 8.

References

A. Beardon [1983], *The Geometry of Discrete Groups*, Springer-Verlag.

P. de Fermat [1640], Letter to Mersenne, 25 December 1640, *Œuvres*, vol.2, 212.

J. Gray [1986], *Linear Differential Equations and Group Theory from Riemann to Poincaré*, Birkhäuser.

J.L. Lagrange [1773], Recherches d'arithmétique, *Nouv. mém. de l'acad. sci. Berlin*, 265ff., Also in his *Œuvres*, vol.3, 695–795.

H. Poincaré [1960], Mathematical Creation, in *The World of Mathematics*, vol. 4, (Ed. James R. Newman), George Allen & Unwin Ltd., London, 2041–2050.

H. Poincaré [1985], *Papers on Fuchsian Functions*, Springer-Verlag.

W. F. Reynolds [1993], Hyperbolic geometry on a hyperboloid, *Amer. Math. Monthly*, **100**, 442–455.

P. J. Ryan [1986], *Euclidean and Non-euclidean Geometry: An Analytical Approach*, Cambridge University Press.

W. Scharlau and H. Opolka [1984], *From Fermat to Minkowski*, Springer-Verlag.

H. A. Schwarz [1872], Über diejenigen Fälle, in welchen die Gaußische hypergeometrische Reihe eine algebraische Function ihres vierten elementes darstellt, *J. reine angew. Math.* **75**, 292–335.

Theory of Fuchsian Groups

by Henri Poincaré

Acta Mathematica I (1882), 1–62 (Sections I and II)

In a series of memoirs presented to the Academy of Sciences, I have defined certain new functions which I call Fuchsian, Kleinian, *theta*-Fuchsian and *zeta*-Fuchsian. Just as the elliptic and abelian functions permit the integration of algebraic differentials, the new transcendants permit the integration of linear differential equations with algebraic coefficients. I have succinctly summarised the results obtained in a note in the *Mathematische Annalen*. Intending now to give a detailed exposition, I begin, in the present work, by studying the properties of Fuchsian groups, reserving until later their consequences from the point of view of function theory.

I. Real transformations[70]

Let z be a complex variable defined by the position of a point in the plane; let t be a complex function of this variable defined by the relation

$$t = \frac{az+b}{cz+d}. \tag{1}$$

I assume, without loss of generality, that we have

$$ad - bc = 1.$$

If the point z describes two arcs which cut at a certain angle α, then the point t for its part will describe two arcs which cut at the same angle α, i.e.

[70] Translator's Note. Poincaré calls them *substitutions*.

the transformation $\left(z, \frac{az+b}{cz+d}\right)$ preserves angles.[71]

The function $\frac{az+b}{cz+d}$ is indeed monogenic.

If z describes a circle, then t also describes a circle; i.e., the transformation $\left(z, \frac{az+b}{cz+d}\right)$ changes circles into circles.

Finally, if z_1, z_2, z_3, z_4 are four values of z and if t_1, t_2, t_3, t_4 are the corresponding values of t, then[72]

$$\frac{t_1 - t_2}{t_1 - t_3} \frac{t_4 - t_3}{t_4 - t_2} = \frac{z_1 - z_2}{z_1 - z_3} \frac{z_4 - z_3}{z_4 - z_2}. \tag{2}$$

There are in general two values of z which equal the corresponding values of t; we call them the *fixed points*[73] of the transformation (1).

If

$$(a+d)^2 \gtrless 4,$$

the fixed points are distinct; and if we call them α and β the relation (1) can be written

$$\frac{t-\alpha}{t-\beta} = K \frac{z-\alpha}{z-\beta} \tag{3}$$

where K is a constant I call the *multiplier.*

If on the other hand

$$a + d = \pm 2$$

the fixed points coincide and we have $\alpha = \beta$.

Relation (1) can then be written

$$\frac{1}{t-\alpha} = \frac{1}{z-\alpha} \pm C. \tag{4}$$

These are the main properties of the linear transformations $\left(z, \frac{az+b}{cz+d}\right)$. However, we have to make one more assumption; we assume that the coefficients

[71] In what follows I use the notation of Jordan. The transformation $[z, f(z)]$ or $x, y; f(x,y), \varphi(x,y)]$ is the operation which changes z to $f(z)$, or that which changes x to $f(x,y)$ and y to $\varphi(x,y)$. The inverse transformation to $[z(f(z)]$ is $[f(x), z]$; the product of two transformations is the operation which consists in making the two transformations successively.

A system of transformations forms a *group* if the inverse of each transformation of the system, and the product of any two transformations of the system, is likewise in the system.

A group A is *isomorphic* to a group B for each substitution in B there is a unique substitution in A such that the product of two substitutions in B corresponds to the product of the corresponding substitutions in A.

If B is likewise isomorphic to A the isomorphism is said to be *holoedric,* otherwise *meriedric* (Translator's note. These are our *isomorphism* and *homomorphism.*)

[72] Translator's note. This equation states equality of *cross-ratios,* $(t_1, t_4; t_2, t_3) = (z_1, z_4; z_2, z_3)$.

[73] Translator's note. Poincaré calls them *double points.*

a, b, c, d are real. I then say that the transformation (1) is a *real transformation.*

It follows that the imaginary part of t is positive, zero or negative according as the imaginary part of z is likewise positive, zero or negative; i.e., the transformation (1) preserves the real axis, which I shall henceforth call X, and it likewise maps the half plane[74] above the axis onto itself.

If z describes a circle with its centre on X, t will also describe a circle with its centre on X. If z_1 and z_2 are complex conjugates, then the corresponding values t_1 and t_2 of t will also be complex conjugates.

If α and β are two values of z, if γ and δ are the corresponding values of t, and if α', β', γ', δ' are their conjugates, then we have, by virtue of relation (2), that

$$\frac{\alpha-\alpha'}{\alpha-\beta'}\,\frac{\beta-\beta'}{\beta-\alpha'}=\frac{\gamma-\gamma'}{\gamma-\delta'}\,\frac{\delta-\delta'}{\delta-\gamma'}.$$

If we make the abbreviation

$$\frac{\alpha-\alpha'}{\alpha-\beta'}\,\frac{\beta-\beta'}{\beta-\alpha'}=(\alpha,\beta),$$

then the latter relation can be written

$$(\alpha,\beta)=(\gamma,\delta). \tag{5}$$

The real transformations have been studied by different geometers, in particular by Klein in his researches on modular functions. They are divided into elliptic, parabolic and hyperbolic transformations.

The elliptic transformations are those for which

$$(a+d)^2<4.$$

The fixed points α and β are conjugate imaginaries; hence one of them is above X and the other below. The relation (1) can be put in the form (3) and the constant K is an imaginary or negative quantity whose absolute value is unity. If z describes a circle through α and β, then t likewise describes a circle through α and β, which cuts the first circle at an angle equal to the argument of K.

The transformation (1) leaves invariant each circle whose centre is on the prolongation of $\alpha\beta$ and which cuts the latter segment harmonically.

The parabolic transformations are those for which

$$(a+d)^2=4.$$

The fixed points coincide in a single one, which is situated on X. The relation (1) comes under the form (4), so that α must be real. If z describes

[74]Translator's note. Poincaré calls it the *part of the plane.*

a circle through α, then t likewise describes a circle through α, tangent to the first. The transformation (1) leaves invariant the circles which are tangent to X at α. Let C be such a circle, and m_0 a point on it. The transformation (1) changes m_0 to another point m_1 on the same circle, m_1 to another point m_2 of C, m_2 to another point m_3, etc. As κ tends to infinity, the points m_κ tend to α. Let m_{-1} be the point which the transformation (1) changes to m_0, m_{-2} the point which it changes to m_{-1}, etc. As κ tends to infinity, the point $m_{-\kappa}$ also tends to α.

I let C_κ denote the circle with centre on X which passes through α and m_κ, and which consequently cuts the circle C orthogonally at α and m_κ. It is clear that the transformation (1) changes C_{-1} into C_0, C_0 into C_1, C_1 into C_2, etc., and in general C_κ into $C_{\kappa+1}$. Moreover, if κ is positive or negative infinity, C_κ reduces to a circle of infinitely small radius. I.e., if we apply the transformation (1), or its inverse, infinitely many times to a circle through α with its centre on X, the radius of this circle becomes infinitely small.

It follows that an arc of finite length which does not meet X cannot meet infinitely many of the circles C_κ, i.e., the successive transforms of a circle C_0 through α with its centre on X.

The hyperbolic transformations are those for which

$$(a+d)^2 > 4.$$

The fixed points α and β are distinct and situated on X. The relation (1) takes the form (3) with K real and positive. Also I can always assume that $K > 1$. If z describes a circle through α and β, then t likewise describes a circle through α and β, and tangent to the first. The transformation (1) leaves invariant the circles which pass through α and β.

As above, I let C denote a circle through α and β, and let

$$\ldots \quad m_{-2}, \quad m_{-1}, \quad m_0, \quad m_1, \quad m_2, \quad \ldots$$

be a series of points such that the transformation (1) changes m_κ into $m_{\kappa+1}$. It is clear that the point m_κ tends to β as κ tends to $+\infty$, and to α as κ tends to $-\infty$.

I also let C_κ denote the circle, with centre on X, which passes through α and m_κ. As κ tends to $+\infty$, C_κ tends to the circle with diameter $\alpha\beta$; as κ tends to $-\infty$ the radius of C_κ decreases indefinitely, i.e., if we apply the transformation (1) an infinite number of times to a circle C_0 passing through α and with its centre on X, we obtain in the limit a circle with diameter $\alpha\beta$. If we apply the inverse transformation to C_0 an infinite number of times, the limit radius will be zero.

If, on the contrary, we apply the transformation (1) infinitely often to a circle through β and with its centre on X, the limit radius will be zero, while if we apply the inverse transformation to it infinitely often we obtain in the limit the circle with diameter $\alpha\beta$.

It follows that an arc of finite length which does not cut X will meet an infinite number of the circles C_κ, i.e., the successive transforms of the circle C_0, or else a finite number of these transforms, according as it does or does not meet the circle with diameter $\alpha\beta$.

If we have $(\alpha, \beta) = (\gamma, \delta)$ then there is a real transformation[75] which changes α into γ and β into δ. This transformation is defined by the relation

$$\frac{t-\gamma}{t-\delta}\,\frac{\gamma'-\delta}{\gamma'-\gamma} = \frac{z-\alpha}{z-\beta}\,\frac{\alpha'-\beta}{\alpha'-\alpha}.$$

There is another property of real transformations which I want to point out; in differentiating the relation (1) one finds

$$\frac{dt}{dz} = \frac{1}{(cz+d)^2}.$$

Moreover, if I let y be the imaginary part of z, and let Y be the imaginary part of t, then I find

$$\left|\frac{dt}{dz}\right| = \frac{Y}{y}.$$

II. Congruent figures

I say that two figures are *congruent* if one is the transform of the other by a *real* transformation. The real transformations form a group; it is clear that two figures congruent to a third are congruent to each other.

I can immediately state the following theorems:

In two congruent figures, the corresponding angles are equal.

If, in two congruent figures, the point γ corresponds to α and the point δ corresponds to β, then

$$(\alpha, \beta) = (\gamma, \delta) \tag{1}$$

This relation can take another form.

Consider the quantities α, β and their conjugates α', β', so that

$$(\alpha, \beta) = \frac{\alpha-\alpha'}{\alpha-\beta'}\,\frac{\beta-\beta'}{\beta-\alpha'}.$$

[75]It is necessary that α and γ (and consequently β and δ) be on the same side of X; if not, the determinant $ad - bc$ will be negative. N.E.N. (Note by the editor of Poincaré's collected works).

The four points α, β, α', β' are on the same circle with centre on X. Suppose in addition that α and β are both above X. The circle cuts X at two points I call h and k; h will be the one on the arc $\beta\beta'$, and k the one on the arc $\alpha\alpha'$. I put

$$[\alpha, \beta] = \frac{\alpha - h}{\alpha - k}\frac{\beta - k}{\beta - h}.$$

$[\alpha, \beta]$ is necessarily real, positive and greater than 1. Moreover,

$$(\alpha, \beta) = \frac{4[\alpha, \beta]}{([\alpha, \beta] + 1)^2}.$$

If γ is a point on the arc $\alpha\beta$ of the circle, we have

$$[\alpha, \gamma][\gamma, \beta] = [\alpha, \beta].$$

It is now clear that, by employing this new notation, we can put the relation (1) in the form

$$[\alpha, \beta] = [\gamma, \delta].$$

We look at what happens when α and β become infinitely close.

Let

$$z = x + y\sqrt{-1},$$
$$dz = dx + dy\sqrt{-1},$$
$$|dz| = \sqrt{dx^2 + dy^2}.$$

Neglecting infinitely small terms of higher order, we have[76]

$$[z, z + dz] = 1 + \frac{|dz|}{y}$$

or

$$\log[z, z + dz] = \frac{|dz|}{y}.$$

Thus we see that the napierian logarithm of $[z, z+dz]$ is proportional to the absolute value of dz and independent of its argument.

The integral

$$\int \frac{|dz|}{y}$$

along an arbitrary curve will be called the L of that curve.[77]

The double integral

$$\int\int \frac{dxdy}{y^2}$$

[76]Translator's note. In the second formula below Poincaré uses L for log.

[77]Translator's note. This is prompted by Poincaré's use of L for log a few lines above.

taken over the interior of any plane region, will be called the S of that region.

It follows from the preceding that two congruent arcs have the same L and two congruent areas have the same S. The L of the arc $\alpha\beta$ of a circle with centre on X will be the napierian logarithm of $[\alpha, \beta]$.

I cannot pass in silence the connection between the preceding notions and the non-Euclidean geometry of Lobachevsky.

Suppose that we agree to strip the words *line, length, distance, area* of their usual meaning, and let a line be any circle with the centre on X, let the length of a curve be what we have called its L, let the distance between two points be the L of the arc of the circle which joins the two points and has its centre on X, and finally let the area of a plane region be what we have called its S.

Suppose moreover that we let the words *angle* and *circle* retain their usual meaning, but agree to call the *centre of a circle* the point which is at a constant distance (in the new sense of the word) from all points of the circle, and let the radius of the circle be this constant distance.

If we adopt this terminology, *the theorems of Lobachevsky are true*, i.e., the new quantities satisfy all the theorems of ordinary geometry, except those which depend on Euclid's postulate.

This terminology has been of great service in my researches, but to avoid any confusion, I do not employ it here.

Memoir on Kleinian Groups

by Henri Poincaré

Acta Mathematica III (1882), 49–92 (Section I)

I. Complex transformations

In a previous memoir[78] I have studied the discontinuous groups formed by linear transformations with real coefficients. In the present work I intend to expose certain results concerning groups of linear transformations with complex coefficients. These transformations are divided naturally into four categories, as we shall see.

Let

$$\left(z, \frac{\alpha z + \beta}{\gamma z + \delta}\right)$$

be any transformation, where I always suppose that

$$\alpha\delta - \beta\gamma = 1.$$

If we have

$$(\alpha + \delta)^2 = 4$$

then the transformation can be put in the form

$$\left(\frac{1}{z-a}, \frac{1}{z-a} + K\right)$$

where a and K are constants. We say that it is *parabolic*.

If we have

$$(\alpha + \delta)^2 \gtrless 4,$$

[78] *Theory of Fuchsian groups* N.E.N. (Translator's note. Footnote by editor of Poincaré's *Œuvres*.)

then the transformation can be put in the form

$$\left(\frac{z-a}{z-b},\ K\frac{z-a}{z-b}\right),$$

where a, b, K are constants.

If we have

$$(\alpha+\delta)^2 \text{ real positive and } > 4,$$

then K is real positive and the transformation is *hyperbolic.*

If we have

$$(\alpha+\delta)^2 \text{ real positive and } < 4,$$

then K is imaginary or negative and of absolute value 1; the transformation is *elliptic.*

Finally, if $(\alpha+\delta)^2$ is imaginary or negative, then K is likewise imaginary or negative and the transformation is *loxodromic.*

A common property of all linear transformations is that they transform circles into circles. Indeed, if we follow the example of Hermite and represent the complex conjugates of quantities u, v, ... by u_0, v_0, ..., then the transformation

$$\left(z,\ \frac{\alpha z+\beta}{\gamma z+\delta}\right) \tag{1}$$

can be written

$$\left(z_0.\ \frac{\alpha_0 z_0+\beta_0}{\gamma_0 z_0+\delta_0}\right).$$

Moreover, the general equation of a circle can be written

$$Azz_0 + Bz + B_0 z_0 + C = 0 \tag{2}$$

where A and C are necessarily real[79] and it is clear that (1)[80] returns us to the circle (2) from the circle whose equation is

$$\begin{aligned} &A(\alpha z+\beta)(\alpha_0 z_0+\beta_0) + B(\alpha z+\beta)(\gamma_0 z_0+\delta_0)\\ &\quad + B_0(\gamma z+\delta)(\alpha_0 z_0+\beta_0) + C(\gamma z+\delta)(\gamma_0 z_0+\delta_0) = 0 \end{aligned}$$

or

$$\begin{aligned} zz_0&(A\alpha\alpha_0 + B\alpha\gamma_0 + B_0\gamma\alpha_0 + C\gamma\gamma_0)\\ &+ z(A\alpha\beta_0 + B\alpha\delta_0 + B_0\gamma\beta_0 + C\gamma\delta_0)\\ &+ z_0(A\beta\alpha_0 + B\beta\gamma_0 + B_0\delta\alpha_0 + C\delta\gamma_0)\\ &+ (A\beta\beta_0 + B\beta\delta_0 + B_0\delta\beta_0 + C\delta\delta_0) = 0 \end{aligned} \tag{3}$$

[79] For the radius of the circle to be real, it is also necessary that $BB_0 - AC > 0$. One easily checks that the analogous expression formed from the coefficients of equation (3) is $(BB_0 - AC)(\alpha\delta - \beta\gamma)(\alpha_0\delta_0 - \beta_0\gamma_0) = BB_0 - AC$ and consequently positive. N.E.N.

[80] Translator's note. Poincaré views this as the coordinate change which renames z as $\frac{\alpha z+\beta}{\gamma z+\delta}$.

We now write down the equation for inversion in the circle (2), i.e., the operation which changes any point z into its transform by reciprocal radius vectors, taking the pole of the transformation to be the centre of the circle (2) and the parameter of the transformation to be the square of the radius of this circle. Here is that equation:

Letting t be the transform of z under the inversion, we have

$$t + \frac{B_0}{A} = \frac{BB_0 - AC}{A(Az_0 + B)}.$$

Let C_1 and C_2 be any two circles, and let I_1 and I_2 denote the respective inversions in these two circles. If we subject an arbitrary point to the inversion I_1, followed by the inversion I_2, then we can represent the resulting operation by the notation $I_1 I_2$. It will be a linear transformation, as one easily convinces oneself. This transformation will be parabolic if the two circles C_1 and C_2 are tangential, elliptic if they cross, and hyperbolic if they do not meet.

Suppose that we observe the resultant of not just two, but of several successive inversions. If the number of inversions is even, then the result will be a linear transformation; if the number is odd, the result will be a more complex operation which can be regarded as a linear transformation followed by an inversion. Moreover, *each linear transformation can be regarded as the resultant of an even number of inversions in infinitely many ways.*

Thus the group obtained by combining the different imaginable inversions in various ways contains all linear transformations.

We put

$$z = \xi + \eta\sqrt{-1}$$

where ξ and η are the coordinates of a point in the plane representing z.

Now consider an arbitrary point, not in the $\xi\eta$ plane, but in space, and let ξ, η, ζ be its coordinates. I assume ζ is positive, so the point considered is above the $\xi\eta$ plane. We have seen that the transformation (1) changes an arbitrary point ξ, η of the ξ, η plane to another point of the same plane. We are going to extend the definition of the transformation (1) in such a way that it applies, not only to a point of the $\xi\eta$ plane, but to any point of space. As we have seen, the transformation (1) can be viewed as the result of a certain number of successive inversions with respect to certain circles in the $\xi\eta$ plane which I call $C_1, C_2, \dots, C_n$. Let $\Sigma_1, \Sigma_2, \dots, \Sigma_n$ be the spheres which have the same centres and the same radii as these circles. Consider the operation which consists in effecting n successive inversions with respect to the spheres $\Sigma_1, \Sigma_2, \dots, \Sigma_n$. This operation, if applied to a point of the $\xi\eta$ plane, does not differ from the transformation (1). Thus we have redefined the transformation (1) so as to make it applicable to a point in space outside the $\xi\eta$ plane.

An inversion with respect to one of the spheres $\Sigma_1, \Sigma_2, \ldots, \Sigma_n$ transforms spheres into spheres, transforms the $\xi\eta$ plane into itself, preserves angles, and transforms each infinitesimally small figure into a similar infinitesimally small figure. Thus all these properties hold for the resultant, i.e., for the generalised transformation (1).

Suppose that an inversion with respect to the circle C_1, for example, changes a certain circle K of the $\xi\eta$ plane into another circle K_1 of the same plane. If S and S_1 are the spheres with the same centres and radii as K and K_1, it is clear that the inversion with respect to the sphere Σ_1 will change S into S_1. Then if the transformation (1) changes the circle K into a certain circle K_n, and if S and S_n are the spheres with the same centres and radii as K and K_n, the generalised transformation (1) will change S into S_n. In fact, the transformation (1) is equivalent to n inversions with respect to the circles $C_1, C_2, \ldots, C_n$ respectively; these inversions change the circle K into K_1, then $K_2, \ldots$, then finally into K_n. If S_i is the sphere with the same centre and radius as K_i the generalised transformation (1) will be equivalent to n inversions with respect to spheres $\Sigma_1, \Sigma_2, \ldots, \Sigma_n$ respectively, and these inversions will change the sphere S successively into $S_1, S_2, \ldots$, and finally S_n. Q.E.D.

To justify the preceding definition, it is necessary to establish the following:

The transformation (1) can be regarded *in infinitely many ways* as the result of an even number of inversions with respect to circles in the $\xi\eta$ plane. Thus the circles $C_1, C_2, \ldots, C_n$ are not perfectly determinate. It is necessary to see that the generalised transformation (1) is nevertheless a perfectly determinate operation. Suppose in fact that the transformation (1) can be regarded:

1° On the one hand, as the resultant of n inversions with respect to the circles $C_1, C_2, \ldots, C_n$;

2° On the other hand, as the resultant of p inversions with respect to p other circles $C'_1, C'_2, \ldots, C'_p$ in the $\xi\eta$ plane.

Let Σ'_i be the sphere with the same centre and radius as C'_i.

Now let P be any point of space. If we apply to it the inversions in the spheres $\Sigma_1, \Sigma_2, \ldots, \Sigma_n$ successively, then we obtain a certain point Q. If we apply to P the inversions in the spheres $\Sigma'_1, \Sigma'_2, \ldots, \Sigma'_p$ successively, then we obtain a certain point Q'. The two operations, which we call (P, Q), (P, Q'), both satisfy the definition of the generalised transformation (1), hence it is necessary to show that the two points Q and Q' coincide. Well, we can pass three spheres S, S', S'' through the point P, with their centres on the $\xi\eta$ plane and cutting it in three great circles K, K', K''.

The transformation (1) will change these three great circles into three other circles K_1, K_1', K_1'' in the $\xi\eta$ plane. Let S_1, S_1', S_1'' be the spheres with the same centres and radii as these circles; the operation (P, Q), and likewise the operation (P, Q'), will change S, S', S'' into S_1, S_1', S_1''. The point Q, and likewise the point Q', is then found as the intersection of the three spheres S_1, S_1', S_1''. Thus these two points coincide. Hence the generalised transformation (1) is a perfectly determinate operation. Q.E.D.

It remains to find the equations of this operation. To define a point P of space, we employ the following three coordinates

$$z = \xi + i\eta, \quad z_0 = \xi - i\eta, \quad \rho^2 = \xi^2 + \eta^2 + \zeta^2 = zz_0 + \zeta^2.$$

Since ζ is assumed positive, these three coordinates completely suffice to determine the point P.

Let ρ'^2, z' and z_0' be the three coordinates of the point Q to which P is sent by the generalised transformation (1). The occurrence of P on the sphere with the same centre and radius as the circle (3) is expressed by

$$\begin{aligned} &\rho^2(A\alpha\alpha_0 + B\alpha\gamma_0 + B_0\gamma\alpha_0 + C\gamma\gamma_0) \\ &\quad + z(A\alpha\beta_0 + B\alpha\delta_0 + B_0\gamma\beta_0 + C\gamma\delta_0) \\ &\quad + z_0(A\beta\alpha_0 + B\beta\gamma_0 + B_0\delta\alpha_0 + C\delta\gamma_0) \\ &\quad + (A\beta\beta_0 + B\beta\delta_0 + B_0\delta\beta_0 + C\delta\delta_0) = 0. \end{aligned} \tag{4}$$

If the point P is on the sphere with the same centre and radius as the circle (3), then the point Q must be found on the sphere with the same centre and radius as the circle (2) obtained from (3) by the transformation (1); whence the equation

$$A\rho'^2 + Bz' + B_0z_0' + C = 0. \tag{5}$$

If we regard A, B, B_0 and C as unknowns, these two equations must be equivalent. We therefore have

$$\begin{aligned} \rho'^2 &= \frac{\rho^2\alpha\alpha_0 + z\alpha\beta_0 + z_0\beta\alpha_0 + \beta\beta_0}{\rho^2\gamma\gamma_0 + z\gamma\delta_0 + z_0\delta\gamma_0 + \delta\delta_0} \\ z' &= \frac{\rho^2\alpha\gamma_0 + z\alpha\delta_0 + z_0\beta\gamma_0 + \beta\delta_0}{\rho^2\gamma\gamma_0 + z\gamma\delta_0 + z_0\delta\gamma_0 + \delta\delta_0} \\ z_0' &= \frac{\rho^2\gamma\alpha_0 + z\gamma\beta_0 + z_0\delta\alpha_0 + \delta\beta_0}{\rho^2\gamma\gamma_0 + z\gamma\delta_0 + z_0\delta\gamma_0 + \delta\delta_0}. \end{aligned}$$

These are the equations of the generalised transformation (1).

We now move on to the general properties of this transformation.

If the transformation (1) is elliptic then it leaves unchanged all points of the circle C which passes through the two fixed points, has its centre in the middle of the line joining the two fixed points, and its plane perpendicular to the $\xi\eta$ plane. It likewise leaves invariant infinitely many circles whose characteristic property is that the spheres passing through them are orthogonal to the circle C just defined. We call C the *fixed circle* of the elliptic transformation.

If the transformation (1) is hyperbolic, then there are only two points of the space which are not changed by the transformation: they are the two fixed points in the $\xi\eta$ plane. The transformation leaves invariant all the spheres which pass through these two points.

If the transformation (1) is parabolic, then there is only a single point left unchanged by the operation; moreover, all the spheres which are tangential, at the fixed point, to a certain line in the $\xi\eta$ plane are left invariant.

Finally, suppose that the transformation (1) is loxodromic; it leaves invariant the circle C whose diameter is the line adjoining the fixed points and whose plane is normal to the $\xi\eta$ plane. However, with the exception of the fixed points, it changes all the points of this circle.

In summary, the only transformations which have fixed points outside the $\xi\eta$ plane are elliptic.

We have previously seen that the generalised transformation (1) transforms each infinitely small figure into a similar infinitely small figure. Let us find the ratio of similitude. If we consider a single inversion, then it is evident that the homologous dimensions of two infinitely small figures, one of which is a transform of the other, will be in the same ratio as the ζ coordinates of the centres of gravity of these figures. It will then be the same when we consider the resultant of several inversions instead of one, i.e., when we consider the generalised transformation (1).

Thus if we let ds, dw, dv respectively denote an infinitely small arc, area, or volume, if ζ is the distance of that arc, area, or volume from the $\xi\eta$ plane, if we let ds', dw', dv' denote their respective transforms under the generalised transformation (1), and if ζ' is the distance of ds', dw' or dv' from the $\xi\eta$ plane we have

$$\frac{ds}{\zeta} = \frac{ds'}{\zeta'}, \quad \frac{dw}{\zeta^2} = \frac{dw'}{\zeta'^2}, \quad \frac{dv}{\zeta^3} = \frac{dv'}{\zeta'^3}. \tag{6}$$

We say that two figures are *congruent* when one is the transform of the other under a generalised transformation (1).

We call the integral

$$\int \frac{ds}{\zeta}$$

taken over an arc, the L of that arc. We call the integral

$$\int \frac{dw}{\zeta^2}$$

over a plane or curved region, the S of that region, and finally we call the integral

$$\int \frac{dv}{\zeta^3}$$

taken over a solid, the V of that solid.

It follows from the equations (6) that two congruent arcs have the same L, that two congruent surfaces have the same S, and that two congruent solids have the same V.

Now suppose that we take away the usual meaning of the words *line* and *plane* and let *lines* and *planes* be circles and spheres which cut the $\xi\eta$ plane orthogonally. Suppose also that we take away the usual meanings of *length, area,* and *volume* and let them mean L, S and V. Finally, suppose that we retain the usual meanings of *circle, sphere* and *angle.* We then recognise that, under this interpretation, all the theorems of Lobachevsky's non-Euclidean geometry, i.e., the geometry which does not admit Euclid's parallel postulate, are perfectly exact. We see also that we have a connection between the theory of linear transformations and non-Euclidean geometry. It is the same connection that Klein has used to find all the groups of finite order contained in the linear group.

On the applications of noneuclidean geometry to the theory of quadratic forms

by Henri Poincaré
***Association francaise pour l'avancement des Sciences*, 10th Session,**

16 April 1881, 132–138

A long time ago, Hermite showed that an indefinite ternary quadratic form with integer coefficients is invariant under infinitely many linear transformations whose coefficients are likewise integers. Since not all the properties of these transformations are yet known, I believe it may be useful to point out some that strike me as curious. As my point of departure, I take the important memoirs of Hermite and Selling on this question (Crelle's *Journal*, vols. XLVII and LXXVIII).[81] I begin by recalling the results obtained by these two distinguished geometers, but I shall express them in a slightly different form, more suited to my purpose.

Let F be an indefinite ternary quadratic form. We can write

$$F = (ax + by + cz)^2 + (a'x + b'y + c'z)^2 - (a''x + b''y + c''z)^2.$$

We put

$$\xi = ax + by + cz, \quad \eta = a'x + b'y + c'z, \quad \zeta = a''x + b''y + c''z.$$
$$F = \xi^2 + \eta^2 - \zeta^2;$$
$$X = \tfrac{\xi}{\zeta+1}, \quad Y = \tfrac{\eta}{\zeta+1}, \quad t = X + iY.$$

[81]Hermite's memoirs are from 1854 (*Œuvres*, vol. 1, p. 191 and 200). Before this, Hermite had been concerned with *definite ternary forms*, following ideas of Gauss (Crelle's *Journal*, vol. XL, 1850; *Œuvres*, vol.1, p. 94.)

The memoir of Selling is from 1874. It studies binary and ternary forms, definite and indefinite, and is a systematic development of the methods of Gauss and Hermite. (A. C.) (Translator's note. Another footnote by an editor of Poincaré's *Œuvres*.)

Suppose the form F is invariant under a linear transformation with integer coefficients, that is, the substitution

$$(1)\qquad \begin{cases} x = Ax' + By' + Cz', \\ y = A_1x' + B_1y + C_1z', \\ z = A_2x' + B_2y' + C_2z' \end{cases}$$

gives

$$F = (ax' + by' + cz')^2 + (a'x' + b'y' + c'z')^2 - (a''x' + b''y' + c''z')^2.$$

Then we put

$$\xi' = ax' + by' + cz', \quad \eta' = a'x' + b'y' + c'z', \quad \zeta' = a''x' + b''y' + c''z'.$$
$$F = \xi'^2 + \eta'^2 - \zeta'^2;$$
$$X' = \tfrac{\xi'}{\zeta'+1}, \quad Y' = \tfrac{\eta'}{\zeta'+1}, \quad t' = X' + iY'.$$

I suppose that we have

$$\xi^2 + \eta^2 - \zeta^2 = -1,$$

whence

$$\xi'^2 + \eta'^2 - \zeta'^2 = -1,$$

where ξ', η', ζ' are given as functions of ξ, η, ζ by equations of the form

$$(2)\qquad \begin{cases} \xi' = \alpha\xi + \beta\eta + \gamma\zeta, \\ \eta' = \alpha'\xi + \beta'\eta + \gamma'\zeta, \\ \zeta' = \alpha''\xi + \beta''\eta + \gamma''\zeta, \end{cases}$$

where α, β, γ etc. are real constants such that[82]

$$(2')\qquad \begin{cases} \alpha^2 + \alpha'^2 - \alpha''^2 = 1, & \beta^2 + \beta'^2 - \beta''^2 = 1, & \gamma^2 + \gamma'^2 - \gamma''^2 = -1; \\ \beta\gamma + \beta'\gamma' - \beta''\gamma'' = 0, & \alpha\gamma + \alpha'\gamma' - \alpha''\gamma'' = 0, & \alpha\beta + \alpha'\beta' - \alpha''\beta'' = 0. \end{cases}$$

Moreover, between t' and t there is a relation of the form

$$t' = \frac{ht + k}{h't + k'},$$

where h, k, h' k' are real constants.[83]

[82] It is assumed implicitly that the matrix of coefficients $A, B, \ldots, C_2$ is nonsingular (of nonzero determinant). See below, where this matrix (or substitution) is interpreted as a *noneuclidean displacement.* (A. C.)

[83] Translator's note. For the equivalence between the relations (2′) and the homographic transformation, A. C. refers to a note on p. 280 of vol. 5 of Poincaré's *Œuvres*. The gist of this has been covered in my introduction.

One knows the coefficients of the relations (1) when one knows those of (2); we therefore confine ourselves to the latter.

Here is how one must proceed to find all the reduced forms of F. Let

$$\xi_1, \quad \eta_1, \quad \zeta_1,$$

be three quantities such that

$$\xi_1^2 + \eta_1^2 - \zeta_1^2 = -1, \tag{3}$$

and let

$$X_1 = \frac{\xi_1}{\zeta_1 + 1}, \quad Y_1 = \frac{\eta_1}{\zeta_1 + 1}. \tag{4}$$

One constructs the form

$$\xi^2 + \eta^2 - \zeta^2 + 2(\xi_1\xi + \eta_1\eta - \zeta_1\zeta)^2,$$

which is *definite*, finds the substitution which reduces it, and applies this substitution to the form F (Hermite's memoir in Crelle's *Journal*, vol. XLVII).[84]

Consider a point m_1 in the plane, with coordinates X_1 and Y_1. It will be inside the circle C of radius 1 whose centre is the origin. If X_1 and Y_1 are given, the relations (3) and (4) determine ξ_1, η_1, ζ_1 (which we call the hyperbolic coordinates of the point m_1) and, consequently, the corresponding reduced form.[85] Thus to each point m_1 inside the circle C there corresponds a reduced equivalent of F, and only one. As the point m_1 varies, the reduced form remains the same as long as m_1 stays within a certain region R_0; but it changes once m_1 passes the boundary of this region. The interior of the circle C is thereby divided into an infinity of regions such that the reduced form does not change while the point m_1 remains inside any one of them. But the number of reduced forms is finite;[86] hence there is necessarily an infinity of regions

$$R_0, \quad R_0', \quad R_0'', \quad \ldots$$

[84] 1854. *Œuvres*, vol. 1, p.193. The quantities ξ_1, η_1, ζ_1 are denoted by λ, μ, ν in Hermite's memoir and are the *continuous variables* whose methodical introduction was the subject of an earlier memoir of his (Crelle's *Journal*, vol. LXI, 1850: *Œuvres*, vol. 1, p.164). (A. C.)

[85] It is assumed implicitly that ζ_1 is positive, so the point ξ_1, η_1, ζ_1 is on the upper sheet of the hyperboloid of revolution represented by equation (3). The point m_1 is its projection, on the ξ, ζ plane, through the peak of the lower sheet. The image of the lower sheet will be the exterior of the circle C. This geometric interpretation is explained by Poincaré in the *Analysis* of these *Œuvres*. (A. C.) (Translator's note: See also my introduction.)

[86] This assertion follows from the hypothesis that the coefficients of the indefinite form F in question are integers. This is the essential principle in Hermite's method for the *introduction of continuous variables* (1850, *Œuvres*, vol. 1, p.164). (A. C.)

which correspond to the same reduced form. Let n be the number of distinct reduced forms, and let

$$R_0, \quad R_1, \quad R_2, \quad \ldots, \quad R_{n-1},$$

be a system of n contiguous regions corresponding to the n distinct reduced forms, which it is always possible to find. Let P be the union of these regions. There exists a system of regions

$$R_0', \quad R_1', \quad R_2', \quad \ldots, \quad R_{n-1}',$$

disposed relative to each other as are

$$R_0, \quad R_1, \quad R_2, \quad \ldots, \quad R_{n-1},$$

and corresponding to the same reduced forms. Let P' be the union of these regions; we similarly define P'', P''', ...

We consider any one of these regions, P'' for example. One of the substitutions (2) is such that, when the point m_1 (whose hyperbolic coordinates are ξ_1, η_1, ζ_1) describes the region P, the point whose hyperbolic coordinates are

$$\alpha\xi_1 + \beta\eta_1 + \gamma\zeta_1, \quad \alpha'\xi_1 + \beta'\eta_1 + \gamma'\zeta_1, \quad \alpha''\xi_1 + \beta''\eta_1 + \gamma''\zeta_1,$$

describes the region P''. Moreover, all the substitutions (2) are realised in this way, so that, to study these substitutions it suffices to study the figure formed by the regions P, P', P'', etc.[87]

Now I am going to appeal to noneuclidean, or pseudo-, geometry. I shall write *ps* and *psly* to abbreviate "pseudogeometry" and "pseudogeometrically" respectively.

By a *ps* line I mean any circle cutting the circle C orthogonally. The *ps* distance between two points is half the logarithm of their cross-ratio with the two points in which the *ps* line through them meets the circle C. The *ps* angle between two curves is the usual geometric angle. A *ps* polygon is a portion of the plane bounded by *ps* lines.

Two figures are *psly* equal if there is a system of nine constants

$$\begin{matrix} \alpha & \beta & \gamma \\ \alpha' & \beta' & \gamma' \\ \alpha'' & \beta'' & \gamma'' \end{matrix}$$

such that

$$\alpha^2 + \alpha'^2 - \alpha''^2 = 1, \quad \beta^2 + \beta'^2 - \beta''^2 = 1, \quad \gamma^2 + \gamma'^2 - \gamma''^2 = -1;$$
$$\beta\gamma + \beta'\gamma' - \beta''\gamma'' = 0, \quad \alpha\gamma + \alpha'\gamma' - \alpha''\gamma'' = 0, \quad \alpha\beta + \alpha'\beta' - \alpha''\beta'' = 0;$$

[87] This method amounts to finding the *fundamental region* for the group of substitutions automorphic for the form F, expressed as homographic transformations. (A. C.)

and such that, when the point (ξ_1, η_1, ζ_1) describes the first figure, the point

$$(\alpha\xi_1 + \beta\eta_1 + \gamma\zeta_1, \quad \alpha'\xi_1 + \beta'\eta_1 + \gamma'\zeta_1, \quad \alpha''\xi_1 + \beta''\eta_1 + \gamma''\zeta_1)$$

describes the second figure.[88]

When this is done, we recognise that *ps* distance, *ps* angle, *ps* lines etc. satisfy the theorems of noneuclidean geometry, that is, the theorems of ordinary geometry, except those which are consequences of Euclid's postulate.

It follows from this that the regions P, P', P'', ... are *psly* equal to each other. By a *ps* movement we mean any operation which sends the point with hyperbolic coordinates ξ, η, ζ to a point whose hyperbolic coordinates are linear functions of ξ, η, ζ. This *ps* movement will be a rotation if it has a fixed point, otherwise a translation. A *ps* movement is completely determined when it is known that point a is sent to a_1 and point b to b_1. We then call the movement (aa_1, bb_1). Of course, it is necessary that the *ps* distance between a_1, b_1 equal the *ps* distance between a, b. Two figures are called *psly* equal if we can pass from one to the other by a *ps* movement.

Now suppose that the given form F does not satisfy the conditions in paragraph 299 of the *Disquisitiones Arithmeticæ*, in other words, that it cannot be made to vanish by substituting integers for x, y, z. Then the region P does not extend as far as the boundary circle C. Following its perimeter in the positive sense, suppose that one meets the regions P_1, P_2, ..., P_n in that order. Let b_i be the common border of P and P_i, and let a_i and a_{i+1} be its extremities. The b_i are called the edges of the region P, and the a_i are called its vertices. In following the perimeter, we successively encounter vertex a_1, edge b_1, vertex a_2, edge b_2, ..., vertex a_n, edge b_n, and finally vertex a_{n+1} which is also vertex a_1.

For this reason we say that the edge following vertex a_i is b_i, and the vertex following b_i is a_{i+1}.

---

[88]The *ps* equality defined in this way is a linear transformation, with matrix T, defined by invariance of the quadratic form

$$\psi(x, y, z) = x^2 + y^2 - z^2,$$

or again, by the condition

$$T \times \begin{bmatrix} 1 & 0 & 0 \\ 0 & 1 & 0 \\ 0 & 0 & -1 \end{bmatrix} \times T^* = \begin{bmatrix} 1 & 0 & 0 \\ 0 & 1 & 0 \\ 0 & 0 & -1 \end{bmatrix},$$

where T^* is the transpose of T. It is the form ψ and its polar form which define *ps* distances and angles.

This hyperbolic geometry is also characterised by homographic substitutions of the imaginary parameter t. In the latter form it is studied in Poincaré's celebrated memoir *Théorie des groupes fuchsiens* (§II on *congruent figures*) (A. C.) (Translator's note. The first Poincaré paper in this volume.)

If we join the consecutive vertices of P by *ps* lines, we obtain a *ps* polygon Q.[89] We do the same for the P', P'', ..., so the interior of the circle C is divided into (infinitely many) *ps* polygons Q, Q', Q'', ... These *ps* polygons are *psly* equal, and the *ps* movement which changes P into P', for example, changes Q into Q'. We imagine the *ps* polygon Q, one of its edges a_1a_2, and the adjacent polygon Q_1 along a_1a_2, which corresponds to the region P_1. If we consider the *ps* movement which changes Q into Q_1, then the inverse *ps* movement changes Q into a certain region Q_i adjacent to Q along edge a_ia_{i+1}. There are two possibilities:

Either Q_i is different from Q_1, in which case the edges a_1a_2 are a homologous pair, and the *ps* movement which changes Q into Q_1 changes a_i into a_2 and a_{i+1} into a_1.

Or else Q_i is the same as Q_1, and the *ps* movement which changes Q into Q_1 is a *ps* rotation of 180° about the *ps* midpoint of a_1a_2. Let β be this midpoint. Then we consider it to be a vertex of the polygon Q, so the polygon has consecutive edges $a_1\beta$, βa_1 meeting at angle 180°. These two edges are homologous, they form a pair, and the *ps* movement which changes Q into Q_1 changes a_1 into a_2 and β into β.

Then, thanks to these conventions:

1° The edges of the polygon Q are partitioned into homologous pairs.

2° Each *ps* movement which changes Q into an adjacent polygon changes an edge into the homologous edge.

Once we know the polygon Q and the pairing of its edges, we know all the *ps* movements which change Q into Q', Q'', etc., and consequently P into P', P'', etc. We therefore know all the substitutions (2), and consequently all the substitutions (1).

To fix ideas, suppose we have a quadrilateral $a_1a_2a_3a_4$, where a_1a_2 is homologous to a_2a_3 and a_1a_4 to a_4a_3. The *ps* movements which change Q to Q', Q'', etc. are the resultants of the two movements (a_1a_3, a_2a_2) and (a_1a_3, a_4a_4).

Each property of the substitutions (1) then reduces to a property of the polygon Q. I mention two:

1° Two homologous edges are *psly* equal.

Now consider any vertex, the edge following it, then the homologous edge, the vertex following it, the edge following it, then the homologous

[89] Poincaré is not precise about the conditions for a form to be reduced. In a later memoir (*Œuvres*, vol. 2, p.479) he even says that one can imagine *infinitely many*.

One can ask whether it is possible to choose reduction conditions so that the common border of neighbouring regions P and P' is always a *ps* line. This would enable the region P to be identified with the polygon Q (as is possible for the reduction of decomposable cubic ternary forms) (A. C.)

edge, and so on. In this way we encounter a certain number of vertices before finally returning to the vertex from which we began. The vertices encountered are said to form a *cycle*, and the vertices of Q are partitioned into a certain number of cycles. With this definition:

2° The sum of the angles corresponding to the different vertices of the same cycle is an aliquot part of 2π.

Now suppose that the form F vanishes when x, y, z are replaced by suitable integers. The results are similar, but with a few differences. The region P extends to the circle C. As in the preceding case, we can partition the inside of the circle into infinitely many *ps* polygons Q, Q', Q'', etc., so that the *ps* movements which change P to P', P to P'', etc. change Q to Q', Q to Q'', etc.

However, it can happen that two consecutive edges of the polygon Q do not meet or, if one prefers, they meet at an imaginary vertex.

The edges of the polygon Q are partitioned into pairs, and members of the same pair are *psly* equal.

The vertices of Q are partitioned into cycles as before, but there are two kinds of cycle: the first containing only imaginary vertices, the second containing real vertices.

The sum of angles corresponding to the vertices in a cycle of the second kind is an aliquot part of 2π.

Index